HACKING THE UNDERGROUND

FEMINIST TECHNOSCIENCES

Rebecca Herzig and Banu Subramaniam, Series Editors

Raquel Velho

HACKING THE UNDERGROUND

Disability, Infrastructure, and London's Public Transport System

University of Washington Press / Seattle

Design by Mindy Basinger Hill / Composed in Minion Pro

UNIVERSITY OF WASHINGTON PRESS / uwapress.uw.edu

LIBRARY OF CONGRESS CATALOGING-IN-PUBLICATION DATA

Names: Velho, Raquel, author.

Title: Hacking the Underground : disability, infrastructure, and London's public transport system / Raquel Velho.

Description: Seattle : University of Washington Press, [2023] | Series: Feminist technosciences | Includes bibliographical references and index.

Identifiers: LCCN 2023027629 (print) | LCCN 2023027630 (ebook) | ISBN 9780295751931 (hardcover ; alk. paper) | ISBN 9780295751948 (paperback ; alk. paper) | ISBN 9780295751955 (ebook)

Subjects: LCSH: People with disabilities—Transportation—England—London. | Local transit—Barrier-free design—England—London. | Subways—Social aspects—England—London. | People with disabilities—England—London—Social conditions.

Classification: LCC HV3022 .V45 2023 (print) | LCC HV3022 (ebook) | DDC 362.409421—dc23/eng/20230912

LC record available at https://lccn.loc.gov/2023027629

LC ebook record available at https://lccn.loc.gov/2023027630

♾ This paper meets the requirements of ANSI/NISO Z39.48-1992 (Permanence of Paper).

TO LÉA AND PAULO, *who first took me to London*

Contents

Acknowledgments

The original home of this book is London, where I was first made to feel small. There, I discovered that learning about cities through their public transportation systems teaches you a lot about a place: who is invited in, who is kept out, whose journeys are eased, whose are made tumultuous. In so many ways my experiences in London and with public transportation mirror my experiences with this book: I was made to feel small, my connections weren't always easy, but learning from others was how I made it home. In these acknowledgments I hope to do justice to the places and people that helped me along my travels. Thank you to all named here for smoothing my connections in your own ways.

This book tells stories that were generously shared with me in cafés, markets, offices, private homes, and booked rooms at University College London. I am grateful to every person who spoke to me (in formal interviews or through various social media channels) and use the names many of you chose for yourselves—whether you chose anonymity or full disclosure. Many thanks to Adam, Alan, Alan*, Alanni, Alex Lyons, Alice, Aimee, Andrew, Ann, Anton, Basil, Carl, Char Aznable, Chiara, D, Diana, Faith, a former government officer, Jo, Joker, Kate, Kerstin, Leda, Lianna, Linda, Marie, Marie Claire, Michael J., Peter, Peter*, Robert, Sal, Sophie, Um Hayaa, and Yvonne. A further thank you to Marie, who invited me to an accessibility event she organized. Thank you again to Alan and Yvonne, who invited me to accompany them for a memorable afternoon of traveling through London, and again to Alan, whose openness to continued conversation with me taught me more than this book can earnestly capture. I am grateful to remain in touch with so many of you to this day. Since we met, two among you have become crip ancestors. Your memories hold power and are a blessing.

Thanks are also due to the many disability organizations who spread the word about my work and found it worth sharing with their followers and members. Particular thanks to Transport for All for their work and activism in London and for engaging in conversation with me.

London gave me five years of growth and research. Thank you to the Department of Science and Technology Studies at University College London, where this journey began. I shared my time there with Erman Sözüdoğru, with whom I also shared a home for four years. He is one of the brightest lights in my life. Brian Balmer appears in these pages in numerous hidden ways, not least for being the one who introduced me to the work of Susan Leigh Star. Thank you to Catherine Holloway, in whose lab I first encountered the questions I ultimately grapple with in this book. I am grateful to senior colleagues who imparted and shared so much knowledge, in particular to Chiara Ambrosio, Jack Stilgoe, Joe Cain, Bill Maclehose (you are missed), Simon Werrett. To Emily Dawson and Angharad Beckett of Leeds University, whose kindness in the form of a well-timed donut gave me strength in one of the most defining experiences of my life. You have each shared wisdom in ways that remain with me and that I find myself passing on to others. Special thanks to the colleagues who first taught me that community is the only way through: Elizabeth Jones, Oliver Marsh, Toby Friend, and Julia Sanchez Dorado. Melanie Smallman has remained a dear friend and conference companion. UCL STS also gifted me Tash Cutts and Sadie Harrison, and what gifts they are.

The London chapter ended to bring me to Rensselaer Polytechnic Institute. I see in these pages work that would never have come to fruition if not for the communities (academic, musical, social) I am privileged to be a member of in upstate New York, which have each shaped some part of this work. The Department of Science and Technology Studies at Rensselaer has experienced so much change in the past years that I barely recognize it as what I joined in 2017. A huge thanks to Nancy Campbell, Abby Kinchy, and Jim Malazita for their collective persistence and support. Further thanks to Jen Cardinal, Brandon Costelloe-Kuehn, Guy Schaffer, Chris Tozzi; it's a pleasure to be in community with you. Thank you to all our graduate students, whose passion and curiosity remind me why I became an academic in the first place. My time at Rensselaer has been greatly enhanced by my undergraduate students, especially those in the design, innovation, and society major. I've learned much from each of you and am honored to witness your own journeys. Thank you to Rensselaer's School of Humanities, Arts, and Social Sciences for a FLASH grant that assisted me in this work.

I am grateful for various networks that I've forged in the past years in broader academic worlds. Aimi Hamraie's friendship is a balm. Being with them and Anson Koch-Rein for our biweekly writing sessions was one of the few ways I got this book written at all! I am also grateful to various conversations with

Lee Vinsel (some intellectual, some gossipy). The introduction of this book was presented at various venues that brought about useful feedback and conversation. A big thanks to the STS departments at Virginia Tech and Cornell University (and to Malte Ziewitz in particular) for hosting me in November 2021. I'm also grateful to the Cafézim crowd. Latinx scholars working in the Global North sharing experiences and stories with one another has offered a path toward thriving in the academic context. I give particular thanks to Andrea Ballestero, for assisting me in first organizing this, and Diana Montaño, Fabian Prieto-Ñañez, and Fernanda Rosa as early adopters and supporters of the events. Thanks also to the Tautegory Project, where I have found space to speak freely and collaborate widely. Though I never had the opportunity to meet Susan Leigh Star, I thank her too. Her work provoked in me the desire to think boldly and interconnectedly. It is my most sincere hope that Leigh would have found the ambitions in this book admirable.

This book found a home in the Feminist Technosciences series at the University of Washington Press, only because people there saw my manuscript as holding potential to contribute to conversations I have been reading with appetite for over ten years. Thank you, therefore, to Banu Subramaniam, whom I first met during a cold February conference and who encouraged me to strengthen my manuscript in very important ways. I also thank Rebecca Herzig and my acquisitions editor, Larin McLaughlin, for helping me through this process. Thank you to my reviewers, whose enthusiasm and generous critiques of the book challenged me to further the work in these pages. A huge thank you for the gargantuan task of copyediting to my editor, Laura Keeler, and for her work in preparing the manuscript for publishing to my project editor, Jennifer L. Comeau. In the words of my father when he thanked his editor decades ago, these are the folks who "helped rescue the dignity of this text." Any remaining errors are mine alone.

My life in upstate New York is richer thanks to the musical world of the Capital Region. Thank goodness for you all; thank goodness for music; thank goodness for dancing. Thanking local bands might not be common in academic book acknowledgments, but it feels right. These people provided the soundtrack to writing sessions and to my evenings away from work. Big thanks at the very least to Front Biz, Zan & The Winter Folk, Chris Bassett, Steve Hammond (and his multiple bands), the Abyssmals, Sophia Subbayya Vastek and Organ Colossal, Pony in the Pancake, Battleaxxx . . . This list is in absolutely no way finite, but let the record show my gratitude for our community. Thank you to Jonah

Moberg and Connor Armbruster for our brief but glorious time as Joan Kelsey's Silver Lining.

My gratitude also goes to the friendships outside of the academic world that held me steady. You are too many to name, but you know yourselves. Zan Strumfeld, in particular, is to be thanked for leading me to open my cello case after ten years of it being metaphorically closed—thank you for holding my hand through so many moments in Troy. Thanks to Laura Rabinow, David Bissember, and sweet Devi for being steadfast friends. The COVID-19 pandemic was only survivable because of board game sessions and backyard picnics with my "pod": James Searle, Jackie Hayes, Peter Lavery, Steph Collins, and Lour Aguas. Thank you for the company in the most unprecedented times. I absolutely cannot fail to name James Searle again, my book doula (sorry about the commas). The summers of 2021 and 2022 included hours making structural sense of the story I offer here. Thank you for your patience and generosity, James, and for crafting beautiful summer reading groups. Members of those reading groups varied according to their familial duties, interest, and capacity, but I thank Lourdes Aguas, Joseph Henderson, Andries Hiskes, Sam Hushagen, and Devin Short (who read this book in a messy first draft state) for our wonderful conversations.

It is no secret that I found the field of STS through the family I was born into. They continue to support my journey in academia. Thank you, siblings, siblings-in-law, and niblings: Homero and Daniela, Tiago and Valerie, Raphaela, Juliano, Zephyr, Virginia, and Dylan. You've all sustained and nourished me in countless ways. My writing is nowhere near as poetic as my father Paulo's, but any ounce of poiesis contained within is thanks to him, the potter, who molded clay into beauty. Words cannot capture how much I miss him, and how grateful I am for the light his memory offers me. My mother Léa is the reason for my venturing into this field when she took me to Esocite in Buenos Aires in 2010. In truth, she's the reason for most of what I am. What a privilege it is to be her daughter and to learn from her every day. Obrigada, meus amores.

The bulk of this work was done in a home in Albany, New York. I also worked in an apartment in Troy and several cafés, hotel rooms, and offices around the world, but mostly in this home in Albany. I am lucky to share that home with two companions. Ziggy, my companion species, humbles me and my desire for order. And lastly, but absolutely mostly, thank you to my husband, Raurri Jennings, for your endless support and energy. The writing of a book tests anyone's limits, and you endured mine. I want you to know I'm so thankful for the sweet love that you bring.

In Their Words

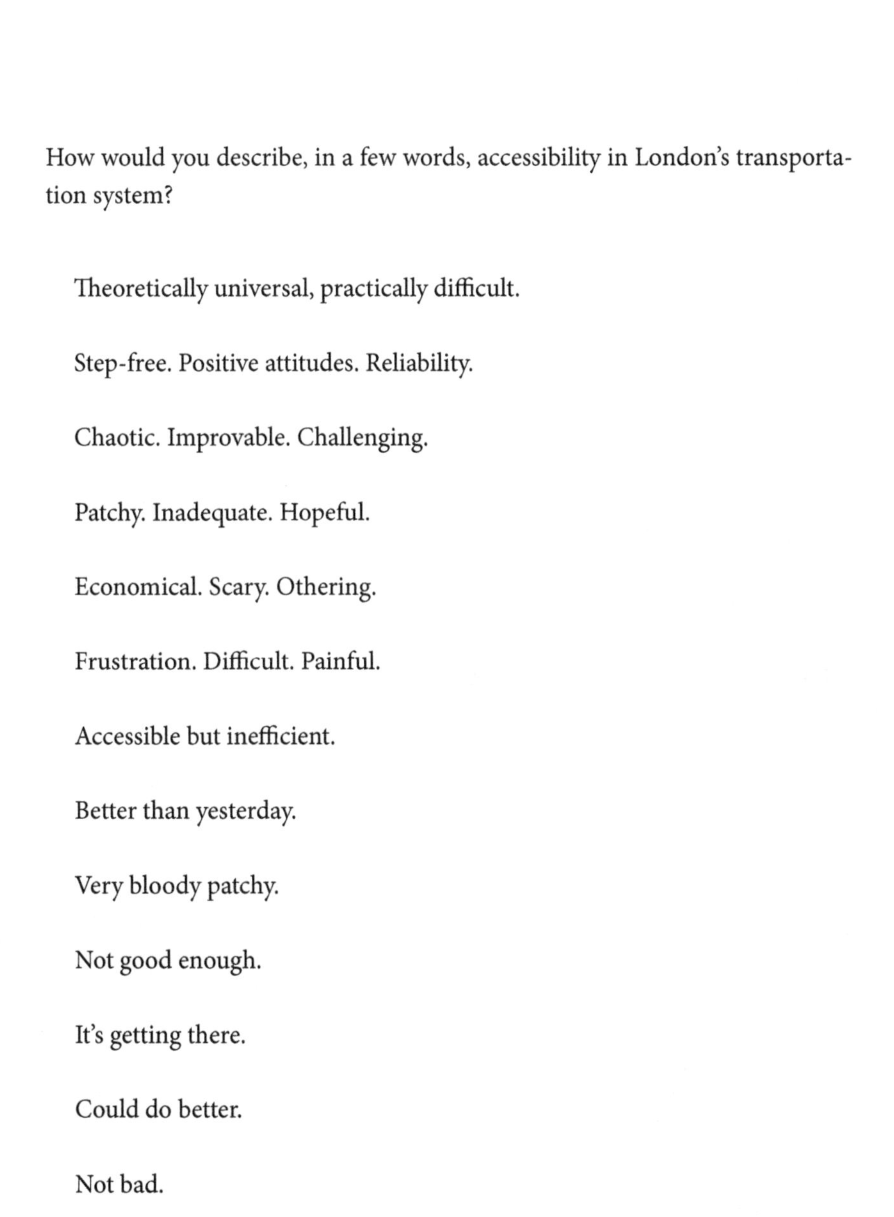

How would you describe, in a few words, accessibility in London's transportation system?

> Theoretically universal, practically difficult.
>
> Step-free. Positive attitudes. Reliability.
>
> Chaotic. Improvable. Challenging.
>
> Patchy. Inadequate. Hopeful.
>
> Economical. Scary. Othering.
>
> Frustration. Difficult. Painful.
>
> Accessible but inefficient.
>
> Better than yesterday.
>
> Very bloody patchy.
>
> Not good enough.
>
> It's getting there.
>
> Could do better.
>
> Not bad.

Could be better.

Behind the times.

Better than it was.

Needs improvement.

Steadily getting better.

Not very accessible yet.

Local knowledge required.

Slow. Unreliable. Frustrating.

Mixed. Varied. Unpredictable.

Labyrinthian. Circuitous. Full of stairs.

Well-intentioned. Inadequate. Uninspiring.

Improving. Better than anywhere else. Inaccessible.

HACKING THE UNDERGROUND

Introduction

One day, early in the fall, I set out from my apartment in East London and headed west to Richmond Station. It was a trip that took about an hour, unremarkable by London standards, and two modes of transport were involved: the Overground and the Underground. At the ticket hall of the station I met Alan and his partner, Yvonne. Alan, a wheelchair user, had invited me to spend an afternoon traveling through London with them. At an interview a few weeks before our journey he had told me about how he had done these trips with a variety of people in the past—managers at transportation companies, politicians, and others. He spoke to me of how important these trips were. He planned out a route for us on which, he said, I would be able to witness the accessibility (or lack thereof) of all modes of transport in London. As he put it, "Because unless you've been and done it, unless you've seen it, it can be quite hard to imagine [what it's like]."

We made our way to one of the train platforms in Richmond, where Yvonne unfolded a lightweight ramp Alan carries on the back of his wheelchair. She confidently placed it at the train's doors when it arrived so that Alan could board. "We can't ask for the ramp here," Alan told me after we settled into seats. Though this station does have access ramps at the platforms we were boarding, we had not bothered doing so. We would be alighting to change lines at Hammersmith Station—at a platform not technically deemed accessible. He explained that if they had asked for the station's portable boarding ramp that technically ought to be used when boarding at this station, the employee would then have to ask where we wanted to disembark. If we had said Hammersmith Station, Alan continued, the employee might hesitate and tell them that the Hammersmith platform was not an accessible one—a fact Alan and Yvonne were well aware of, but they also knew that the line exchange was feasible if you just used your own ramp. After all, the next train we were going to board would be arriving on the other side of the very same Hammersmith platform we would be alighting at. Better not to bother the employees with the ins and outs of the plan.

So we didn't. Aboard the train, five stations later, the audio announcement informed us that the next stop would be Hammersmith. Yvonne waited until the train's doors opened to place the ramp again. Alan disembarked; I followed. The buzzer signaling closing doors wailed, but their portable ramp was still wedged between the train and the platform as the doors closed upon it. Alan's partner tugged hard—once, then a second time. My heart raced, fearing their ramp was lost as the train began to move. It finally came loose as Yvonne tugged a third time. She swiftly folded it and tucked it away on the back of Alan's wheelchair. Neither Alan nor Yvonne seemed remotely fazed. Alan continued speaking to me as though nothing remarkable had happened. He pointed out the width of the platform where we had disembarked—how close to the train the pillars were. I shook myself off, processing and taking notes of his remarks. As I followed them to the other side of the platform, I also reflected on our different reactions to this moment: my fear and hesitancy while Alan and Yvonne continued moving forward smoothly. To them, things were going precisely to plan; I was the novice.

It is my hope that this book will offer similar moments of reversal. This is a book about infrastructures, and I consider myself a proud member of the "Society of People Interested in Boring Things" (Star 2002, 157). Following the stories of wheelchair users (such as Alan) navigating the public transportation system in London, I bring together the fields of disability studies and science and technology studies (STS) to call attention to infrastructures in ways that highlight their ironies and tensions. I offer a crip feminist lens, expanded upon below, through which to understand infrastructures and reframe some of the ambiguities that have always been present in discussions about the topic. My primary aim is to show how understanding infrastructures through experiences recaptures and perfectly exemplifies the caveat that Leigh Star and Karen Ruhleder proposed in their 1996 article: "Analytically, infrastructure appears only as a relational property, not as a thing stripped of use" (Star and Ruhleder 1996, 113).

My focus is on analyzing infrastructures from the relations and experiences of groups that were—and in many ways remain—not part of the membership or the community of practice. This methodological focus on the experience of marginalized groups in relation to infrastructures highlights how even seemingly successful networks occupy much more complex states of (non)functionality, refusing the static definitions that have often been attributed to them. As will become apparent in the chapters to come, descriptors such as *functional* and *invisible* and even *deterioration* quickly come apart when scrutinized through the lens of crip feminist technoscience. Instead, in prioritizing the *relational state* of infrastructures in analyses, we can understand all of these terms as possible descriptions of the same infrastructure, for the same users, concurrently.

Studying infrastructures not solely through their history or their internal constitution but through marginalized users' experiences and interactions with them also illustrates how the built environment holds the potential for both inclusionary and exclusionary world building. In this way, I offer a nuanced argument that acknowledges the hegemonic force of infrastructures while highlighting the creative and potentially transformative practices of neglected and marginalized users. In this sense, this book—like much of the scholarship on infrastructures—is about tensions (Star 1999; Howe et al. 2016). If my original ambition when I began this work was to analytically resolve some of these tensions, the crip feminist lens I adopted to analyze the case now urges me to embrace them. I argue that it is through these tensions—temporal, scalar, or otherwise—that infrastructures survive, even if they do not always thrive.

This book also attempts to grapple with a particular dissatisfaction that I have accumulated over years of reading about and studying infrastructures. I could not resolve the tension between two commonly used quotes, "infrastructures are invisible until they break down" and "any genuine infrastructure is invisible," and my interlocutors' descriptions of their experiences with a celebrated infrastructure: public transportation in London. How could a transportation system that so often appears on lists of "top 10 transportation systems around the world" (Falzon 2017; Chenel and Moynihan 2021) be the same one that interviewees described as "pure frustration, really" (Carl) with the potential to "leave like a crack in the heart of the disabled person" (Um Hayaa)?[1] It felt impossible to close the gap between what seemed like different worlds: is the system that enables nearly four billion journeys annually, according to the governmental organization Transport for London (2021)—arguably, therefore, a genuine infrastructure—the same one that wheelchair users described as requiring a "mental travel map" (D) of "safe routes" (Alanni) to get around?[2] If one passenger can complete a journey but another passenger cannot, when do we say the infrastructure is broken and thus no longer invisible?

Some scholars have questioned the concept of invisibility through the infrastructural need for constant repair and maintenance work. In this sense, infrastructures become, instead, mundanely visible (Harvey and Knox 2012; Larkin 2013, 2008; Ureta 2014). However, I have found this binary of invisibility and visibility an unsatisfying duality to navigate, as these distinctions are often made on the basis of breakdown or disrepair. From this definition, scholars have at times drawn a line between infrastructures of the Global North and South. In the latter, some have argued, "norms for infrastructure can be considerably different," with services "routinely failing" (Edwards 2002, 188). Meanwhile, infrastructures of the Global North, where resources are presumably less scarce and politics more transparent, are somehow more invisible. This distinction of northern and southern infrastructures is particularly unsatisfying considering various cases in the Global North where infrastructures are failing specific populations, whether we're discussing accessibility in transportation for disabled persons or access to clean water in majority-Black Flint, Michigan. Infrastructures of the Global North might seem functional and invisible, but the question that often remains is, to whom? London's public transport system would seem to fall into the invisible category: it is a network with significant financial subsidies that enables billions of journeys every year. I dedicate an entire chapter to untangling the meaning of invisibility in its application to infrastructure. I propose that an experiential understanding of invisibility can take us farther in our analyses of infrastructures, not because the idea of invisibility must be salvaged for any particular reason, but because it enables us to think of the particular functions of infrastructures in more nuanced ways.

Ultimately, the purpose of this book is to contend with what seems like a double bind of infrastructures. These sociotechnical systems, as they might also be called, have often provoked questions of scale: how does one situate an analysis of an infrastructure if there is no single unit?[3] As they are described, infrastructures have reach—they telescope out of any single site or event. Further, they are the embodiment of standards that shape and are shaped by conventions of practice, which affect the expectations and demands of how one interacts with them. Despite this, infrastructures are also localized and personal; our interactions with them happen in specific moments and demand sensory engagements. These engagements can, and often do, occur as the scripts demand of us—until they do not. We know that users in all forms can hack, subvert, and rescript their interactions with technologies (Oudshoorn and Pinch 2003). While the literature in the field of science and technology studies offers important insights in how users and nonusers shape technology, those at the margins of use, or excluded from use, are less explicitly discussed. I focus, then, on these spaces of marginality and use them as a point of departure (Star 1991). These high-tension zones offer the opportunity to show that, yes, infrastructures, by their nature, demand and reflect conventions of practice but, in so doing, also create space for multiplicity. As Leigh Star wrote, "those in wheelchairs . . . are good points of departure for our analysis because they remind us that, indeed, it might have been otherwise" (Star 1991, 53). As I will argue, wheelchair users not only remind us that it might have been otherwise; they often make it so.

Before finding my way to these theoretical concerns, however, the original question that guided the work in this book was straightforward: How do wheelchair users use public transportation in London? In the years I lived in the British capital, seldom a month went by without at least one news story on wheelchair users struggling to navigate transportation systems in the country (Coleman 2016; Moss 2013). Most stories emerged in London, perhaps due to London's transportation system being so iconic or the centrality of London in British media. Characterized by the red and blue roundels, bright vermillion double-decker buses, and the quintessential "mind the gap" audio, London public transport is even used in movies and TV shows to telegraph to viewers where the story takes place. London's transit system is also one of the most extensive in the world. The Underground network alone spans 250 miles and 270 stations, with an additional 112 stations as part of the Overground network. London also has a fleet of over 9,300 buses serving 675 routes with more than 19,000 stops. Though impressive, this description offers little as to what navigating this system is like as a wheelchair passenger.

We might begin to reframe the system and understand its accessibility by looking into the number of stations that are defined by Transport for London, the administrative authority, as having "step-free" access. Step-free access is described as "stations [that] have lifts or ramps—or a combination of both—so that customers don't have to use escalators or stairs to move between the street and the platform" (Transport for London n.d.). The term is often, but not exclusively, used as a catchall term for accessibility to the infrastructure, if problematically so. Based on this definition alone we already see the first hints of accessibility concerns in London transport. Someone dependent on step-free stations for navigating the city can access only 91 of the 270 stations on the Underground network and 60 of the 112 ones on the Overground. The observant pedestrian will also catch, at various bus stops, large white posters with red letters stating, "Buggy users, please make space for wheelchair users," a hint of what has become a sustained debate over the dedicated priority area in buses.[4]

While there have been numerous changes to the transportation system in the past thirty years and many strategies are currently in place to meet new accessibility goals, the experience of wheelchair users within this infrastructure is still extremely fraught. This book opens with the voices of wheelchair users interviewed for this project. Take a moment to go back to the page before this introduction and read the brief answers by wheelchair users to the question, "How would you describe accessibility in London's transportation system?" It is a fascinating collection of answers. Some interviewees opted for adjectives or nouns describing qualities, others for brief sentences. The answers are often contradictory or counterintuitive: accessible but inefficient, hopeful but frustrating, improving but inadequate. These contradictions, even in such short answers, already tease at a more interesting story than figures alone could supply. This is a story of frustrations and innovations, exclusion and inclusion.

Thus, the work presented here is centered in the particularities of the case, exploring wheelchair users' experiences with London's public transport system. It provides a two-part answer to the question, How do wheelchair users use public transport in London? I argue that wheelchair users use public transportation in London with significant difficulties, having to counter various barriers embedded in the system that are the result of this infrastructure having consolidated in an era when the marginalization of disabled persons was par for the course. This history and the ensuing barriers and exclusion are discussed, respectively, in chapters 1 and 2. The second part of the answer is that despite these difficulties, wheelchair users do use public transportation in London. They do so through the development of *belligerent techne*, which I define as strategies and tactics of subversion and control (see chapters 2 and 3). Consequently, the work of this marginalized group results in an active shaping of the infrastructure—a shaping that is not accounted for in past histories of London transport, nor in infrastructure studies.

I understand that the work presented here is of limited scope as compared to broader conversations of accessibility for disabled persons other than wheelchair users. Indeed, readers will rarely encounter mentions of assistive devices besides wheelchairs. While some might see this as a potential weakness to the arguments of the book, I believe that my focus on wheelchair user accessibility is a strength. Because wheelchair users are more easily socially understood and perceived as "disabled" than various other persons, particularly those who negotiate access with less apparent mobility issues, accessibility for these passengers is often articulated as being resolved through the provision of what, in chapter 3, I call *access icons*: ramps, elevators, and wheelchair priority areas. As I will argue, these access icons often fall short of "simply" providing access. The theory of access used in this book must therefore account for this irony, and must be broader and more complex. I understand accessibility to be a relational category of movement that enables one person's entry and exit from built environments. That entry or exit not only is supported through alleged access icons but also is an entanglement of sociocultural and material stuff that needs to be differently negotiated depending on a person's positionality, needs, capacities, and abilities. Indeed, as we will see, what the infrastructure offers as access icons can themselves create more complicated permutations of access friction, including the new passenger debates between wheelchair users and caretakers with baby carriages mentioned above.

In answering such a straightforward question as "how do wheelchair users use public transportation in London?" I offer contributions to how to study infrastructures as well as to our specific understandings of the concept. For the former, I developed an analytical lens that enabled me to capture the hegemonic character of infrastructures without losing the experiences of marginalized users in their interactions with them. I offer this lens as a crip feminist approach to infrastructure studies, which I describe below. Regarding the latter, I make two specific contributions to past conversations on infrastructure studies that are dialectic in nature. Specifically, I argue that marginalized groups play a significant role in shaping infrastructures that surround us, often in ways that members of the community of practice may not notice but do benefit from. However, the shaping done to infrastructures is often limited by retrofitting. In using the term *retrofitting*, we close off possible future imaginations as we continue to uphold the original normative functions of our infrastructures.

A CRIP FEMINIST LENS FOR INFRASTRUCTURES

First let me expand on the conceptual and analytical lens that also grounds this book's methodological approach. I described this framework as a *crip feminist lens*. I understand this to be a lens that builds off the long history of feminist technoscience scholarship by bringing to it the work of scholars in the field of critical disability studies. The lens is particularly indebted to scholarship that already lies at the intersections of disability studies and feminist technoscience, such as the excellent special issue of the journal *Catalyst: Feminism, Theory, Technoscience*, edited by Kelly Fritsch, Aimi Hamraie, Mara Mills, and David Serlin (Fritsch et al. 2019). The crip feminist lens is further supported and inspired by methodological and analytical approaches such as emancipatory action research that have informed the way that interlocutor quotes are woven into my book as points of focus. In sum, I propose the crip feminist lens for infrastructure studies as a way to keep a necessary double vision: on the one hand, seeing and exposing the way things are and, on the other, seeing how they are also already otherwise (and could lead down even newer paths). This double vision extends in many other ways, including the ability to contend with both bottom-up and top-down discourses and in being dedicated to ecological systems of thinking.

To propose a crip feminist technoscience requires active engagement with the "non-innocence" of science and technology while not eschewing them entirely—thus recognizing how our lives are also deeply entangled with these non-innocent artifacts (Haraway 1991). In this sense, the analysis here is necessarily informed by the work of feminist technoscience scholars, many of whom reject dichotomies and dualisms and instead privilege fractured identities. Feminist technoscience enables here a critical stance that allows for embracing ironies and partialities. Feminist technoscience is imbued with the understanding that perspectives and, importantly, knowledges are only ever partial, contextual, and therefore situated to a person's position (Haraway 1988; Harding 2016).

Feminist technoscience literature is one in which tensions and dichotomies are found more often than not. It enables a double vision that, on the one hand, recognizes how technoscience has been, and is still, the result of privileged partial perspectives (many of them harmful, with origins in the military-industrial complex), while also holding promissory power for the betterment and liberation of subjugated lives. More recently, feminist technoscience has taken the shape of "xenofeminism" (Cuboniks 2018), an ambitious manifesto along the lines of the cyborg, that strives toward learning from the local, applying emancipatory tactics at a larger scale. Xenofeminism proposes itself as a platform with a "mutable architecture" that "remains available for perpetual modification and enhancement following the navigational impulse of militant ethical reasoning" (Cuboniks 2018, 59). The goal is to learn from ethical adversaries, observe how stability and order seem to emerge from spontaneity, and "widen our aperture of freedom" through hacking existing platforms (Cuboniks 2018, 65). Ultimately, this is a normative commitment: that our worlds are unjust and require changing. The slogan remains: "If nature is unjust, change nature!" (Cuboniks 2018, 93).

While these literatures push the conceptual framework to allow tensions and ironies to emerge (and, crucially, remain) in analysis, what is commonly cited as feminist technoscience offers little in terms of tools for analyzing infrastructures. Here, I turn to the rich interdisciplinary literature on infrastructures, which has spanned anthropology, sociology, geography, and history of technology, among many other fields. Leigh Star and her various collaborators are key contributors to the lens I propose, and in many ways, I see in her work a clear articulation of feminist technoscience as well. The more I have engaged with Star's work over the years, the clearer it is that though she is often placed in the realm of infrastructure studies, she ought also to be among the scholars we cite when speaking of feminist technoscience, though her work is often absent from those collections. I have found in her work the key elements for analyzing infrastructures, from proposed ethnographic approaches to beautifully refined definitions. I have also found in her scholarship a profound care and commitment to pluralism, irony, and even the humor we are so often used to encountering in feminist technoscience. I am fairly certain that Star would place her work in the realm of feminist theory, and her closest colleagues and friends have certainly done so (Clarke 2010). In a working paper published in the edited collection *Boundary Objects and Beyond: Working with Leigh Star* we see some of the sharper edges of Star's feminist theory and how it is woven seamlessly into her work. In one piece, titled "Misplaced Concretism and Concrete Situations: Feminism, Method, and Information Technology," Star argues for a feminist theory and method that places multiplicity at the heart of analysis, drawing from the work of her mentor, Mary Daly, as well as that of bell hooks, Gloria Anzaldúa (from whom she takes her concern for the borderlands), and Patti Lather (Star 2015). We can certainly see this concern and approach through much of her work.

I fold Star's work into my scholarship in part because of her dedication to thinking well beyond classic disciplinary lines. Star seemed comfortable speaking to diverse audiences, be it in giving interviews to *Technical Communication Quarterly* about her contributions to computer science (Zachry 2008) or writing in *Systems Practice* about an "unnameable" transdisciplinary group of scholars whose work straddles similar analyses of the situated, distributed, and material forms of knowledge and cognition (Star 1992). In this sense, Star more than just espoused "ecological thinking"; she embodied it. She avoided functional or functionalist approaches and favored ecological analyses that, in her terms, are meant to treat "a situation in its entirety looking for relationships," understanding the web of relations to be seamless—a "network-without-voids" (Star 1995, 15, 27). In her writing she put this ecological approach into practice as she wove through her own ecological relations, deftly citing cognitive scientists, computer scientists, and philosophers. Her scholarship is more than a sum of each of these ways of thinking. Rather, she composed compelling narratives that synthesized the contributions of Deweyan pragmatism with situated and distributed insights on cognition to argue for research that opens "into the world of work and practice—which require oxymorons, heterogeneity, and unexpected juxtapositions of the formal and empirical, the distributed and the local" (Star 1992, 407).

Star brought this approach to the heart of her work in infrastructures as she and coauthors began looking into the processes through which systems consolidate. Their particular concern has tended toward the history of standards, classification, and categories, particularly the consequences of these standards. Following the work of her teacher, Anselm Strauss, Star was dedicated to "stud[ying] the unstudied" through the aforementioned ecological understandings, never hiding the "social justice agenda" that was underpinned by "valorizing previously neglected people and things" (Star 1999, 379). Star long argued for the need to identify master narratives, showing them for what they are, denaturalizing them in the process, and thereby naming the unnamed "others."

Of particular importance to the crip feminist lens for infrastructures is the emphasis that Star always placed on communities of practice, borderlands, and relationality. Always the ecological thinker, Star argued that these concepts need to be thought of interconnectedly. To be a member of a community of practice is to be in a particular set of relationships with objects that pertain to that community. "Membership," she wrote, "can thus be described individually as the *experience* of encountering objects, and increasingly being in a naturalized relationship with them" (Star 2015, 154, original emphasis). Thus the borderlands—the margins, the monsters, those both in and out or neither in nor out—become the spaces whence one can denaturalize an infrastructure. But the borderlands also are the spaces where anomalies, resistances, and suffering exist. As María Puig de la Bellacasa puts it, Star's dedication to studying the borderlands was "not only about pain, violence, and survival: these fissures are also about possibility" (de la Bellacasa 2015, 49). I attempt to envelop a similar dedication in this lens, which captures not only how wheelchair users navigate transportation but how, in so doing, they forge new realities.

Here, this book also takes from and contributes to user studies within STS. User studies, from the late 1980s to the early years of the twenty-first century, wanted to break out of a linear understanding of how technologies develop to show how users affect the shape of artifacts (Oudshoorn and Pinch, 2003). Chapters in *How Users Matter: The Co-Construction of Users and Technologies* provide us with an alternative narrative and an ideal starting point for attempting to understand how particular demographics of users, not just the designers, shape technologies and their uses. My research bridges some of the gaps between these literatures by investigating a user group that is less often investigated, misfit users, as work has focused on groups on either side of the divide (users or nonusers; insiders or outsiders). In *How Users Matter*, for example, Sally Wyatt addresses the importance of nonusers and their choice of not using a technology as not being related to deviance or inequality but as a rational choice (Wyatt 2003). In the case of my own research, it will be interesting to investigate the case of (arguably) excluded users and their impact—not those who choose not to, but those who are restricted in their uses. This high-tension zone so interests me because it is unsatisfied by an analysis that might otherwise be content with describing circumstances as complex as infrastructures and their users as either *users* or *others*, *users* or *nonusers*. The elegance of Star's work has always been, to me, the emphasis and care placed on analyzing our relationships with technology as necessarily coming from multiple memberships. Through the multiplicity of norms and standards, we may all fall short of always meeting all expectations. After all, as Goffman reminds us, "In an important sense there is only one complete unblushing male in America: a young, married, white, urban, northern, heterosexual Protestant father of college education, fully employed, of good complexion, weight and height and a recent record in sports" (Goffman 1963, 128). Fall short of one, and your own identity may be spoiled in some respect in your interactions.

Despite Star's care for multiplicity and the borderlands and her recurrent use of disabled persons and access as examples of the brick walls of standards (Star 1991; Star and Ruhleder 1996), she never truly tackled the topic of infrastructures from the perspective of disability. Thankfully, significant work has been done in the field of disability studies that bolsters the work I am endeavoring to do here, some of which expands on the feminist technoscience literature and some of which carries important genealogies of its own. I thus turn to the work of scholars such as Alison Kafer, Rosemarie Garland-Thomson, and Robert McRuer.

Of key concern in the crip feminist lens for infrastructure studies are the mirror concepts of the normate and the misfit, both developed by Garland-Thomson. The concept of normate is articulated in Garland-Thomson's book *Extraordinary Bodies*, where she, too, quotes Goffman's wry sentence, noting his description as the "normate position." But the concept of the normate is more than this; it is a conceptual subject position against which disability is measured and marked. Indeed, the normate "designates the social figure through which people can represent themselves as definitive human beings," an imagined "everyman" against which disability becomes a marker of stigma (Garland-Thomson 1997,

8). Standing in some regards as its mirror image, the term *misfit* was developed in relation to theories of feminist materialisms, which have also shaped my scholarship. Garland-Thomson proposes misfit as an "incongruent relationship," the attempt to make two things come together when the context does not easily allow it. It evokes material as well as relational concerns of spatial juxtaposition. In her words, "misfits are inherently unstable rather than fixed," and thus the concept serves to "theorize disability as a way of being in an environment, as a material arrangement" (Garland-Thomson 2011, 594). It is dependent on material forms as well as on the relation among material forms in context. These concepts are particularly useful in my work as they interrelate with ideas of marginalized users, whom I prefer instead to call *misfitting users*, because they are not what infrastructures understand as the imagined and therefore normate user.

Work from critical disability studies as it intersects with science and technology studies, such as that of scholars Alison Kafer, Aimi Hamraie, and Kelly Fritsch, are also present in this crip feminist lens. In these literatures, the understanding of technoscience and the promises of STS as theory are questioned through showing how technoscience is not done from the top down to disabled people. Rather, scholars demonstrate the ways that technoscience is produced by disabled people. This work further adds dimensions that were, frankly, unexplored in feminist technoscience, though disabled people are, in those theories, often upheld as the best examples of cyborgs. In problematizing this near fetishization of disabled people as cyborgs, Kafer crips the cyborg to show that it is not sufficient to uphold disabled people as the "best" examples of the cyborg merely because of their dependence on technology. Rather, Kafer emphasizes, cripping the cyborg highlights the necessarily political nature of cyborgs. Cyborgness leans on acts of defiant, severe, complex negotiations with technologies (and beyond), not on a "need" for technology (Kafer 2013). In this process cyborg theory gains a new robustness that goes beyond "required" technological or nonhuman associations for survival and passes through political acts and negotiations. In the same move, disability studies gains an analytical tool; one useful in thinking about complex relationships with technologies that aren't reduced to ableist notions of dependency. After all, as Sara Hendren reminds us, all technology is assistive (Hendren 2014).

The work of scholars at the intersections of STS and critical disability studies, such as Aimi Hamraie and Ashley Shew, adds a powerful layer to this lens that not only enhances the question of the politics of cyborgness. Their work reminds us that the ways that disabled people interact with technology are not determined solely by external authorities. Classic discourses of technology as simply liberatory or empowering for disabled people are mired in what Shew identifies as *technoableism* (Shew 2020). This tendency to see in technoscience ways through which disabled people can fully integrate society tend to valorize specific understandings of the body. Yet while this is happening, we are also reminded that disabled people are producers of technoscience on their own terms. The concept of crip technoscience holds us to centering disabled people as sources of knowledge and creation of technoscience. This concept is refined in Hamraie and Fritsch's *Crip Technoscience Manifesto*, where they define crip technoscience as "a field of relations, knowledges, and practices that enables the flourishing of crip ways of producing and engaging the material world" (Hamraie and Fritsch 2019, 4). The four commitments of crip technoscience they offer center the work of disabled people as knowers and makers, enhancing the ways in which access is a collective political experience, not an individual one based on individual impairment. This brings into focus the various "relational circuits" that disabled people weave among bodies, environments, and tools, a tapestry of interdependence that is central to survival and the emergence of better worlds.

This leads me to the methodologies that have guided and inspired the work here. I at times understand my interest in emancipatory action research as directly related to my Latin American roots and the work of Orlando Fals Borda and Paulo Freire. In Latin American STS, many questions are based on critical approaches to the question of social inclusion. What is social inclusion? Inclusion in what, of whom, through what process, on whose terms? Reading Freire and Fals Borda, each of whose work on liberation and emancipation discusses oppression as a process that can (and should) be countered, it became clear to me that to be a researcher is to wield tremendous social capital. Choices are presented: to paint oneself as a modest witness or to pick up the tools of one's profession to "advance people's struggles for power and justice" (Fals Borda 1996, 71). Fals Borda was a proponent of action research and, within disability studies, scholars such as Mike Oliver have made similar calls for emancipatory research. Working and writing at a time when disabled people had little to gain by participating in research, Oliver realized that their needs and struggles were often going unheard. Disabled persons participating in research were reduced to their impairments without full consideration for their personhood and experiences. Disabled scholars worried about the reproduction of power relations within research (Oliver and Barnes 2006). Mike Oliver thus called for emancipatory research on disability so as to "illuminate the lived experiences of progressive social groups"; in order to do so, "[research] must also be illuminated by their struggles" (Oliver 1992, 107). Oliver's emancipatory research prizes recognition of participants in research not as fragments but as individuals with knowledge and experience. Being inspired by disability studies scholars allowed me to gain confidence in my approach and provided guidance in establishing my credibility with interviewees to earn their trust in my research. As a result, the guiding question of this research also owes to this methodology, particularly as outlined by Mike Oliver. It focuses on describing experiences while redefining the "problem." Rather than asking why some wheelchair users do not use public transport, it asks how those who do use it manage to do so.

In sum, the crip feminist lens on infrastructure studies is particularly indebted to these four branches of scholarly work: feminist technoscience, infrastructure studies at the borderlands, critical disability studies, and emancipatory research methodologies. It has resulted in what I see as three key analytical commitments. First, feminist technoscience holds me to not resolving tensions and ironies; these are desirable outcomes of a double vision that enables the privileging of partial perspectives "from below" while also seeing a larger picture of hegemony. Second, thanks to critical disability studies and crip technoscience, this lens aims to recognize and identify the knowledge and labor of marginalized positions while seeing the borderlands not solely as spaces of suffering but as spaces of active production. This is particularly true for the knowledge and power of disabled people. Last, the goal of using the crip feminist lens is to redistribute the results and recommendations that emerge from research back to the communities and stakeholders that collaborated with or have power within the scope of the project.[5]

In this way, the crip feminist lens for infrastructure studies can take many shapes. For me, in this book, it has offered a way to identify the practices and experiences of wheelchair users in public transport in London as world-building and infrastructure-shaping endeavors that inform other relations within the system as well. For others, a crip feminist lens for infrastructures may help unearth a plethora of ways in which our infrastructures include notions of normate users, causing previously unknown misfits to come to the fore. Ultimately, if the use of the crip feminist lens for infrastructures enables us to see the experiences of new misfits, it may also highlight the plurality of knowledges and practices that these misfits bring with them. This is the hopeful first step toward enfolding new knowledges into processes that will forge new possibilities.

THE PROMISES AND LIMITS OF SHAPING FROM THE MARGINS

Based on the crip feminist lens for studying infrastructures, this book makes two key contributions to conversations in infrastructure studies that each chapter builds toward while making smaller theoretical insights. First, the use of a crip feminist lens enables the identification of how those at the margins, our misfits—in this case, disabled passengers—actively shape infrastructures from the outside in. I am not speaking here about simple acts of subversion (though those, too, are present) or ways of skillfully coping with exclusionary systems. My emphasis is, very precisely, on how wheelchair users, through tactical subversions and strategic contestation, mold transportation infrastructure in London. Ultimately, I will show how knowledge from the margins shapes our infrastructures. This is a powerful lesson to learn when we have grown accustomed to hegemonic narratives of the top-down nature of infrastructures. I wish to make it clear that this is not just a question of subversive tactics used in the moment. Rather, these can be durable acts of shaping that affect our infrastructures—with varying timescales of durability. Furthermore, I argue that the result of this shaping means that the experiences of other users, the original community of practice of an infrastructure, are shifted and even mediated by users in the margins. This opens space for various other multiplicities and debates. An example of how experiences are shaped by the work and knowledge of wheelchair users is that low-floor buses with wheelchair priority spaces became the new bus standard in London from the early years of the twenty-first century onward. These changes have given parents and guardians with strollers better access to buses as well. A child no longer needs to be lifted out of the stroller, the stroller collapsed and carried up the stairs. This illustrates how wheelchair users' and disabled persons' demands for access have shaped other populations' experiences with transportation as well.

The story would be too simple if it ended there. I further contribute an understanding of why the process of molding from the margins is so limited in large consolidated systems. As I argue in chapter 3, despite the shaping and molding by passengers using wheelchairs, their impact on the infrastructure is still curbed and successes are fraught. This is due to the historical path of the system—the generative entrenchments inherited over time—as well as the imaginations and expectations of its decision makers and community of practice. Infrastructural futures and potentials are thus limited by our historical understanding of what an infrastructure's function is and whom it ought to serve. At this juncture, the idea of retrofitting is necessarily limited by the historical designed materiality of the infrastructure. As I expand and discuss in chapter 2, this in part explains the limitations even with the changes and successes accumulated in transportation infrastructure over the years, such as the rise of the debates over wheelchair priority areas. New imaginations and more knowledge need to be incorporated as legitimate to the infrastructure for it to break out of its original path of momentum and thus begin another. Otherwise, as entrenched values continue to hold their grasp on an infrastructure, successes will continue to be curbed, as they have been with the priority areas. As I will show, this debate is in no small part due to entrenched social norms regarding which bodies are seen as (re) productive and thus welcome in London's transportation system.

BOOK STRUCTURE

My two-part argument unfolds in this book through four thematic sub-arguments that relate the specifics of my case back to the world of infrastructures more broadly. The goal of each chapter is thus not only to tell a part of the answer to my guiding question (How do wheelchair users user public transport in London?), but also to recapture key themes in the literature on infrastructures, offering theoretical expansions on previous work or gently nudging us to dig deeper into ideas that have come to be accepted as given. Thus, the chapters in this book weave between the empirical contents of my research while always returning to theoretical reflections on the topic of infrastructures and experience.

This book is based on fieldwork conducted over two years in London, the majority of it as interviews that took place between 2015 and 2016. The majority of interviewees were wheelchair users. Among the non–wheelchair users were partners who accompanied them and opted to participate (often due to being their traveling companions), but I also interviewed engineers, policy makers, and accessibility consultants at Transport for London and tendered private companies. Their perspectives were useful for discussing the changes that they have seen or implemented in the transportation industry. However, the most significant part of the work here is based on interviews with wheelchair users, whose experiences were the key concern. I am extremely grateful for their generous contributions to this work, as they often spent upward of an hour or two with me. Our interviews were loosely structured. Gender representation was roughly equal among interviewees, with a slight overrepresentation of women and lesser representation of gender nonbinary individuals.[6] Age representation was skewed toward those under sixty years of age (no individuals under eighteen were interviewed), but there was a roughly equal distribution between eighteen and sixty. Information about individuals' impairments was not collected; the need for disclosing diagnoses was deemed unnecessary as interviewees could speak about their personal experiences and capacity without disclosing such personal information.[7] There was also a significant variety in the type of wheelchairs used by interlocutors. Electrical power chairs, manual wheelchairs, and power-assisted chairs are the primary options. In some cases, the interviewees mentioned having more than one type. How these choices can impact experiences in transportation surfaces as an important consideration in chapter 3.

The names used to refer to participants here are either pseudonyms or their own given names. The option to select a pseudonym was offered but, in some cases, participants opted for their own names, which gave me pause in considering questions of confidentiality. There has been some research discussing that "a research participant might want to receive recognition for some or all of what he or she contributes" (Kaiser 2009, 1638). I thus concluded that respect for autonomous decisions should be given to interlocutors. I clarified their confidentiality choices with the interviewees and, as Giordano and colleagues (2007) recommend, discussed that the use of their interview data may not always be what they envision. They still chose to use their own names and to identify themselves; this choice has been respected.[8] Extensive use of interview quotes will be present throughout this book, centering the voices of wheelchair users in this work.

During the two years of fieldwork there were also ample opportunities for ethnographic observations, which served as supplementary data. These observations took place in various settings: disability roadshows, wheelchair skills training sessions, observing or traveling with wheelchair users on public transport, and other transportation events that were held in London in the years of fieldwork.[9] These observation opportunities enhanced the analysis and were further supplemented by extensive documentary collection. The latter corpus of data was mostly used in providing contextual information, particularly concerning the legislative rights of disabled people and the ways that institutions self-define their roles and responsibilities. As such, the documentary collection served as both resource and topic of inquiry: it provided information about the subject but, because collected documents were a production of particular stakeholder groups, they are also a topic of investigation, as they present and represent particular interests and definitions (Zimmerman and Pollner 1970). The bulk of this documentation is official records, largely produced by the English government and its departments and committees. Examples are acts of Parliament pertaining to the rights of disabled people, such as the Disability Discrimination Act of 1995 (DDA95) and the Equality Act of 2010 (EA10). Select committee inquiries from both houses of Parliament were also used, such as the House of Commons Select Committee on Transport's inquiry into access to transport for disabled people or the House of Lords Equality Act of 2010 and its disability committee's report. Other official records include strategy and implementation frameworks published by the mayor of London, a collection of publications by the Department for Transport (DfT), papers and assessments concerning the construction of Crossrail (now known as the Elizabeth line), and some internal documents produced by Transport for London (TfL) for its employees (such as the guide for bus drivers, the Big Red Book). Other documents are commercial media accounts, which encompass newspaper articles on disabled people's access to transport, including particular cases of law suits against transport providers. Further documents are public-facing websites, such as TfL's main website and its various subpages on accessibility and transport, as well as its media-facing press office. The website of the charity Transport for All was also captured, particularly the pages describing the services it provides and its news and blog subsections. Public blogs written by wheelchair users on their experiences with transportation were also used, such as Alan's, titled *Never a Dull Journey*.

Each chapter is titled after a key concern or characteristic of infrastructures that has captured scholarly interest over the years. The chapters are topically centered while further developing the dialectic argument of the book on the promises and limits of shaping infrastructures with knowledge from the margins. Each chapter title is further enhanced by a quote from an interlocutor who spoke, in their own way, to the theme of interest.

Chapter 1, on history, begins with a quote from Carl, who tells us, "Obviously it is a Victorian network." I focus on the tensions of historicity and the diverse values given to different histories. This chapter provides a novel intersection between the history of transport in London and the history of disabled persons in the United Kingdom to illustrate how infrastructures are historically contingent and our current understandings of users, nonusers, and marginalized users are therefore dependent on the past. It is by overlaying these two histories that we can see how the development of public transport in London happened at a time when disabled people were socially segregated and had already been distanced from society for some time, perceived as passive or unproductive. Consequentially, transportation, which had largely developed to serve a compulsorily able-bodied population (McRuer 2006), catering to productive laborers and consumers, had not embedded disabled people's access needs within it. This history is deeply felt by present-day wheelchair users, who often allude to the age of the system as a partial explanation for their exclusion.

A full discussion of the process of shaping infrastructure and its limits takes place in chapter 2, where I articulate the concept of *designed materiality*. Tired of the apologetic responses to his complaint emails, Anton says, "Stop sending me apologies. I want change." I define the concept of designed materiality to capture the differentially entangled histories and orderings of the materiality of infrastructures while still contending with how the form and shape of these infrastructures are the conduit through which this multiplicity occurs. I trace the ways that the designed materiality of transportation has been molded over the years by disabled activists and passengers. I borrow a concept from biology, *niche construction*, as a fruitful metaphor to describe how changes to the infrastructure can be said to occur in this bottom-up way. As I argue that wheelchair users have been gradually and persistently shaping the transportation infrastructure over the past decades, niche construction enables me to show how these actions are curbed and limited by the niche, too. This limitation can be seen as the result of repetitive feedback loops, including in the use and articulation of retrofits as paths forward. Niche construction, as a metaphor, offers instead the potential promise of feedforward loops for alternative futures. These feedforward loops may enable us to identify more socially just and democratic paths for infrastructural design.

In chapter 3, on knowledge, Alan gives us one of his adages for navigating transportation: "They're in charge, but you're in control." He uses this slogan to remind himself that he may take his time while traveling and not allow the rules of the system to rush him in unwarranted ways. In this chapter, I show the regimes of technoscience that are present in the system and how the knowledge of disabled passengers combats and rebuilds—at times successfully—the knowledge that has been legitimized and materialized as a part of London's public transport system. Here we see how wheelchair users' forms of knowledge are actively shaping the transportation system through what I call *belligerent techne*, truant acts of freedom within a system that otherwise does not encapsulate their needs, abilities or knowledges. In mobilizing belligerent techne, wheelchair users not only go against the rules of system; they forge new paths of access within it. Here we also begin to see the limits of that shaping.

In chapter 4, I ask us to go beyond *invisibility* as a descriptor for infrastructures. We begin with Michael J., who tells us that he "feels invisible still" when he uses public transport in London, and I query what is at the root when we say that infrastructures are invisible until they break down. Pulling from concepts in phenomenology, I trace out the notion of invisibility as related to normative functionality and the many states of tools and tool-being. Showing the various barriers that wheelchair users must navigate to travel using public transport, I push Star and Ruhleder's caveat on infrastructures' relational character to the forefront to show what it implies for the idea of invisibility. As I will argue, invisibility is as much about one's positionality within the system as it is about the system itself.

I offer, in the conclusion, reflections on the contributions this book makes to the field of infrastructure studies and ways that the crip feminist lens can enhance our understanding of the relationality of these systems. I underline the importance of highlighting how marginalized communities and their knowledge practices are always already shaping our world and mediating our experiences thereof. I further offer a path forward in policy making for infrastructural design based on the work of john a. powell on targeted universalism. This is a potential way of integrating diverse perspectives and needs as legitimate forms of knowledge into our infrastructures. These are paths that are no longer satisfied with retrofitting the past to create socially just worlds. It privileges, instead, feedforward loops to depetrify our futures.

ONE / Partial Histories / *"Obviously it is a Victorian network . . ."*

In 2013 Transport for London celebrated the anniversary of the London Underground (also known as the Tube), commemorating 150 years of subterranean railways with a study in contrasts. Original steam locomotives were placed back on the rails for a series of trips, followed by the inauguration of the first electric trains on the tracks (Transport for London 2013).

Public transport in London is one of the city's most recognized emblems. The city's tourist shops are filled with Transport for London logo magnets with "Mind the Gap," "London Underground," or famous station names in blue block letters and plastic models of vermillion double-decker buses. The real buses are peppered across every corner of the English capital, and their history is at least as long as that of their train counterparts. The proud history and symbolic power of London's transportation system has a wide reach; it has featured in iconic movie scenes and has been discussed in fiction and nonfiction alike. Indeed, the London Transport Museum is homed in one of the city's most tourist-friendly and prestigious neighborhoods, right next door to the Royal Opera House in Covent Garden.

The history of London's transportation system, however, is not spoken of only as a glorious past or as something to take pride in. For many of the wheelchair users I spoke with, this history was identified as something to be dealt with, outgrown, or bypassed. Among them was Carl, whom I met on a late September afternoon in West London. About halfway into our conversation, he told me about his side projects, besides his job:

> I've also been doing . . . I've been campaigning for a more accessible Tube network by myself now for the past two years. And a friend of mine that's joined me was a former engineer on the Tube network, so he's got quite a good knowledge and understanding of the Tube.

I asked him to tell me about how he started this type of work. He told me about his frustration with lack of accessibility in London and how it had affected his life, including jobs he had had to turn down due to their locations being inaccessible to him. This despite Carl being a fit white man in his thirties who uses a light manual wheelchair he can adeptly maneuver, including over gaps and bumps along his path.

> I was a bit confused, and I didn't really feel that I had an equal access to the job market because of the way the Tube . . . if you can't get to work, and it wasn't feasible to drive, then I felt I was being put at a significant disadvantage finding work. . . . These two experiences [for job interviews] that I had where I was driving, so I'd go into London and thought, nah, can't use the Tube, not even going to attempt it, I'll drive. Drove, got to the place I needed to be for my job interview and discovered there was no disabled parking, and when there was parking, all the places were taken so I was driving around, driving around, driving around, couldn't park my car, missed my appointment, lost out on the option. And then I had a very similar experience two months later, and kind of thought, two quite big opportunities passed me by, and I attribute it to that, to not being able to access the Tube network. . . . It probably took me two hours in the car as well, ironically. If I was able to take the Tube it would probably have taken me thirty-five minutes, so it's worth not being able to park the car. The journey time would have been significantly reduced.

His campaigning focus, Carl told me, has therefore been the London Underground:

> My sort of biggest issue and sort of being in London is really with the Tube network. Because obviously it is a Victorian network, and obviously back then, people with disabilities weren't really considered important. So, we're trying to work with a really old infrastructure, but at the same time I still think it's too slow.

This articulation of the "Victorianness" of the transportation system having embedded so many of the issues into the network was striking. While various other wheelchair users also discussed the age of stations or of the system itself as an issue, it was often articulated in terms of historical preservation concerns or of Transport for London's reluctance toward change and novelty. Carl's statement was therefore striking in its specific connection between the age of the infrastructure and its originators' attitudes toward disabled persons. It is, as Carl puts it, because of the Tube's origins in the Victorian era that it is so hostile to disabled persons. What, precisely, is so obvious to Carl about the Victorian era's lack of care or concern with disability that has endured to this day and still creates the issues he faces, 150 years later? Through the various mentions of the infrastructure's age in our conversations, it became apparent that the consequences of decisions made as far back as the 1860s are felt by wheelchair users to this day. Infrastructures have the dubious quality of carrying ghosts within them. Whether the ghosts constitute a haunting or a protective presence depends on those who identify the spirits.

This is a situating chapter. It seeks to describe, literally, "the situation," understood here as the context within which wheelchair users' paradoxical experiences with transportation emerge. In other words, I offer a historical account of public transportation in London. As Carl identifies, to do this I must track the situation back to its origins in the Victorian era and therefore also account for Victorian attitudes toward disabled persons. I must thus contend with two histories and place them in conversation: the history of disability on the one hand, and the history of public transport in London on the other. What emerges in the patchworking of these histories, however, is something altogether more useful, as it not only provides context for the reader but also advances a key consideration of the book, or what I name the *historical double bind of infrastructures*. This historical double bind argues that while users' experiences of an infrastructure are rarely accounted for in functionalist histories of infrastructure, such experiences always contain echoes of that very history.

In tracing the situation through the experiences of present-day wheelchair-using passengers, I must therefore attempt to trace the history of London transport with a particular attention to concerns about access and how access considerations are stretched and scoped depending on the historical moment. I will first offer the more commonly told history of this transportation system, the prevailing historical narrative encountered in transport history books. This will serve to familiarize the reader with an environment to which they might otherwise be a stranger.[1] This prevailing narrative follows a pattern familiar to historians of technology, reminiscent of Thomas P. Hughes's now classic work on the development of large technical systems (Hughes 1983). These histories tend to focus on the work of the conceptualizers of the system. If we are to trace Victorian ghosts, then, it is useful to outline the more commonly told version before we identify the silences it has left behind.

However, this story fails to tell us very much about disabled users, or indeed any users, within the system. Users themselves are rarely the focus of histories on the development of infrastructures; that is not their primary goal, but it is one of mine. I will therefore offer more contextualized histories of disability and transport. The goal in patching together these histories is to offer insights that might otherwise be lost in their separateness. If the history of the development of transportation is where we begin to identify the ghosts in the system, in providing more context for the Victorianness present in it through a history of disability, we can trace how and why some of the ghosts have remained. This is the chapter's second goal: discussing how historical hauntings of infrastructure remain. It is this patchwork of histories that will highlight what Carl so brilliantly captured in a single sentence: the Victorian ghosts that have remained in the transportation system continue to haunt wheelchair passengers to this day, despite multiple attempts to exorcise them through retrofits. Delineating the joint histories of transport and disability will show their complex commensurability as echoes of the Victorian era persist into the twenty-first century. The perdurance of stabilized infrastructures comes to the fore, and these historical entrenchments enable us to question the very notion of retrofitting, or retrospectively fixing infrastructures over time.

THE HISTORICAL DOUBLE BIND OF INFRASTRUCTURES

In what follows, I trace the history of transportation in London together with that of disability in the United Kingdom. The latter shows how disability rights activism created a shifting landscape in understandings of access and the public in the context of transportation and enables me to treat more directly the fraught character of accessibility as it contends with Victorian ghosts. The history of public transportation in London is a partial one, largely focused on the system's constitution from the standpoint of its builders. This common history thus struggles to include questions of contestation of use, such as those posed by disability activists. That is not to say that specific contestations are not accounted for in the histories of networks. Indeed, Hughes himself argued through questions of competition and the system's resistance to growth through the concept of *reverse salients* (whether technical or organizational).

Reverse salients are a way of understanding a system within its builders' logic. The concept was conceived in full understanding of its militaristic baggage regarding "advancing battle lines" and the need for "concentrated action if expansion is to proceed" (Hughes 1983, 79). Hence, its deployment is most appropriate to describe a specific moment of a system: a moment when the system builders' vision of the network's future becomes (possibly momentarily) untenable due to the contingencies of its past conceptions. The system builder resolves the tension between the past and the future by developing a response that brings the infrastructure from the past into new contingencies to better reach its established goals. In other words, when the reverse salient is satisfactorily fixed, the system can continue its expansion—until a new reverse salient emerges and the cycle begins anew. It seems that infrastructures are often playing catch-up with themselves as system builders attempt to make them work—a theme that has become, in the past ten years, of great interest to scholars whose focus is maintenance. There, the literature shifts focus away from system builders (often teasingly called *the innovators*) to system maintainers (Graham and Thrift 2007; Mattern 2018; Russell and Vinsel 2018; Vinsel and Russell 2020; Jackson 2015). In attempts to rescue labor that had otherwise been relegated to the margins of infrastructural livelihood, scholars of maintenance highlight that infrastructures would fall into disrepair if it weren't for the work of repairers, maintainers, fixers, hackers, cleaners . . . the list goes on. The work of maintenance, then, is to keep the stability of a network intact in the present so that it can continue to pursue its goals. If not for that, it falls into ruins or rubble and becomes another relic of the past (Velho and Ureta 2019). And while historicity is not always explicitly highlighted as the analytical goal of maintenance studies, it is there when maintenance is described as "the work needed to guarantee [the system's] existence through time" (Denis and Pontille 2015, 339).

The reverse salient is not, however, helpful for dealing with contestation and resistance from those outside of the system who would rather see themselves incorporated as users. While Hughes's concern was largely one of thinking through expansion as an internal goal of networks, this approach offers limited insights into how infrastructures contend with external demands. In this sense, infrastructures unfurl outward, not inward: their purpose isn't to serve their creators but those whom their creators deem to be users. They deal with the world *out there*, the environment, or, as Hughes defines it, "those parts of the world that are not subject to a system's control, but that influence the system" (Hughes 1983, 3).[2] Hughes defines the environment as that which is apart from the systems he defines: it influences the system, but it is somewhat unclear whether Hughes understands it as being, in turn, influenced by it. Infrastructure studies needs to trouble this one-sided perspective of the environment, as the shifting landscape of an infrastructure can also enable new forces to act on it. As I will trace in the history below, this is precisely what happens in the collision of the history of London's public transport with disabled activists in the United Kingdom.

The historical double bind of infrastructures is that it must contend not only with the infrastructure's own internal technical or organizational limitations but also with the limits of vision of the infrastructure in its entanglements with its environment: how its future-oriented goal can become dated and clash with the passage of time and the emergence of new demands, new rights, new voices. The survival of infrastructures over time depends not only on fixing reverse salients to allow system expansion but also on a capacity for retrofitting the system to incorporate new visions of its goal (though, as we will see in the next chapter, these retrofits are themselves limited).

Seeing the histories of transport and disability not as parallel but as diffractive, we can tend to their continuities and their ruptures. The ironies of historicity are not left unremarked by wheelchair passengers of the network. The past, present, and future clash in infrastructures as they contend with the expiration dates of both their internal makeup and their conceptual basis. We can discuss this external historical tension by showing it in two steps: In the first we highlight the gap between the histories—showing how infrastructures are products of their time of conception and, in the case of London's public transport system, one that is rife with particular understandings of who the users were going to be and what (and whom) that excludes in the process. In the second we show the moments of collision and consider the process through which an infrastructure endures beyond its moment of creation and survives contestations.

The gap between the histories discussed here is at its widest in the decades when public transport was being developed in London—the turn of the twentieth century. If we read through the brief history of transportation offered below with an eye for that gap, we can see the values of the Victorian era emerging throughout the system. It was an era of a growing and thriving middle class whose upward mobility was dependent on the rise of the industrial era. Public transit in its moments of creation was intimately linked with concerns of efficiency (to solve gridlock), and the subsequent underground boom and the continual consolidation, often via acquisition, of private rail companies with London buses was also often couched in questions of efficiency and profitability. Even at the network's inception, Pearson, one of the fathers of London's underground railway, argued for the necessity of the underground railway not only to make transportation more efficient but also to liberate the working class by improving mobility and access to jobs. The concern was keenly classed, couched as it was in terms of both the possibility of upward social mobility and a more particular concern of geographical mobility for those of the middle and upward classes (be it for jobs or for leisure).

If transportation was being developed during the era of industrialization, it is important to ask where disabled people fit into that moment. Weaving into this the history of disability, we are reminded that in the late nineteenth to early twentieth centuries, disability was not only medicalized and individualized; it was institutionalized and marginalized. Disabled people were socially distanced. The new machineries of production in the modern era required *l'homme moyen*, the average man, which did not account for perceived deviations, like a disabled body. As McRuer highlights of Harry Braverman's work on the links between capitalism and management science, the work toward efficiency and productivity precluded disability and "played a large part in the emergence of the identity of the able-bodied worker" (McRuer 2006, 88). Through the rise of industrialization and the standardization of workers' bodies, through the shoring up of an ableist society, the gap between our two histories becomes clearer. The access needs of disabled people in transportation modes that aimed to circulate workers and tourists, housewives and consumers, would seem irrelevant. Being disabled largely precluded belonging to the former categories.[3] As disabled people were driven either deeper into the private sphere to be cared for by their families or further out to the fringes of society as they were institutionalized, it seems unlikely that any system builders that a Hughesian historical analysis might highlight would take account of their requirements in public transport.

As Howe and coauthors describe it, "A lesson of infrastructure is that it surfaces the social conditions and times in which it is sited; thus, it demonstrates as much about our historical and cultural attentions in a particular moment and place as it does about the thing itself" (Howe et al. 2016, 552). This being the case, the cultural attentions surrounding the birth and consolidation of transportation in London precluded disabled people from the transport system. Yet that also doesn't tell the whole story, largely due to a shift in disabled people's self-understanding and the rise of social rights movements in the later decades of the twentieth century. What happened in that moment? And if the histories collided and transportation began to include disability access requirements, why do interviewees continually emphasize the age of the network system?

Decisions made within networks produce generative entrenchments, or, otherwise put, decisions that are made early in the infrastructure's existence root themselves deep into the network, "becoming increasingly more difficult to eradicate" (Lampland and Star 2009, 14). And when contestation appears and an infrastructure is called into question, one does not simply rebuild from the ground up a brand-new system. In sum, "infrastructure does not grow de novo" (Star 1999, 382). Though the process might not begin from zero, the emergence of social movements, specifically the disability rights movement, made new demands on public transport in London—demands that the infrastructure became forced to contend with. This is when the infrastructure struggles to keep up with reverse salients not from its own expansionist logic, but from outside of itself, with its environment and changes in society demanding new responses and new commitments.

It would be naive to believe that infrastructures are permanent and immune to the passage of time, as the history of transport in London testifies. Stations are abandoned; others are constructed; new rails are built; others rust through use and exposure to weather and are subsequently replaced or discontinued; changing historical contexts and administrations prioritize trains over buses, or vice versa; and new technologies are applied to (or developed specifically for) the network. From its inception in a Victorian past through its progression into the present, it blunders and manages to adapt, never evading its historicity. Because the system is not malleable enough to be entirely reconceptualized, disabled passengers of public transport suffer from the slow process of modular increments within the network, both aware of the reasons for the sluggish pace of progress but also wanting (and campaigning for) it to be more radical. As Carl put it, "So we're trying to work with a really old infrastructure but at the same time I still think it's too slow."

It is these winding and tortuous histories that follow. Whether we begin with the history of transport or the history of disability, little would change for the argument; they are clashing histories that have influenced one another. I begin with a classic history—the prevailing historical account of transportation in London—to question the silences that emerge in that version. I then shift to the history of disability in the United Kingdom, specifically focusing on Victorian conceptions of disability, to contextualize the silences that emerged before. Finally, I show how, in bringing these two histories together, we can identify a shifting understanding of *access* in the context of transportation as disability activists claim their space in the infrastructure. Ultimately, conceptions of accessibility, as we find them in the current situation of London's transit system, have been widened by disability activists while also being limited in their application due to the entrenchments of the past.

A CLASSIC HISTORY, ABRIDGED

My aim in this (radically) abridged history is to trace out the classic narrative that has been inscribed in many histories of transportation books that focus on how London's various modes of transport became icons entrenched in the capital to this day. Ultimately, however, we will have to ask where the silences in this story emerge.

The Rise of Passenger Transport in London

Classic histories of transport in London almost invariably set the scene just before the Victorian era. The majority opt to tell a story of a metropolis undergoing huge transformations. Britain was among the first countries to experience the demographic transition of, concurrently, an increase in fertility rates and a decrease in mortality rates as well as the Industrial Revolution, factors that provoked two population changes in the United Kingdom: a qualitative change as the population urbanized, and a quantitative boom as the number of inhabitants more than doubled in less than a century. The middle classes experienced a particularly significant increase in numbers, in large part due to the industrialization of manufacturing (Barker and Robbins 1963, 1975; Ashford 2013; Martin 2012a; Pedroche 2013; Wolmar 2004).

London's geographical boundaries expanded as the population began to sprawl beyond its walls. Industry developed and previously residential areas became earmarked as business sites. Significant swaths of the population were displaced, with members of the middle classes relocating just outside the city as the city's poorer populations crowded within it.[4] This meant the beginning of urban sprawl, and the distance between the public and private spheres (work and home) began to grow for these middling classes (Barker and Robbins 1963, 1975).

It is thus solidly in the Victorian era that our story gathers speed. The invention of the steam engine revolutionized transport at the time for both freight and passengers. London, as the financial and industrial center of the British Empire, was among the first to be infected by the steam bug: between 1836 and 1852, five railway termini for long-distance travels were inaugurated. These, combined with the construction of hotels and the establishment of museums and galleries, allowed for an increase in the number of tourists visiting the city, partnered with the growth of an affluent middle class, which could now afford to travel from other towns to the city. London was becoming an increasingly attractive destination for business and pleasure.

Current Londoners may be heartened to learn that the capital's struggles with gridlock and heavy traffic are nothing new; indeed, they've been around since the early nineteenth century. The increase in population and the inflow of tourists filled London's streets to the brim with personal carriages, hackney carriages (taxis), wagons, carts, and—the newest addition to the scene—omnibuses. Omnibuses brought communal transportation to London, bypassing the need to book a carriage in advance. Introduced to London in 1829 by George Shilibeer, omnibuses picked up passengers by the side of the road at advertised locations and at precise times.[5] This business model quickly proliferated with several new competing operators that eventually merged under the name London General Omnibus Company in 1855. By 1858 the company operated sixty-three routes with nearly six hundred vehicles.

Horse-drawn carriages provided little amelioration for what continued to be heavy gridlock in the capital. Parliament saw an imperative to remedy the situation, holding a Select Committee on Metropolitan Communications in 1855 that recommended establishing the Metropolitan Board of Works. This new authority, funded by property taxes, became responsible for infrastructural improvement of all types within London, from sewers to streets to parks. The Metropolitan Board of Works saw the need to connect London's railway termini and financial center to relieve congestion. Among the proposals made to link these termini was the building of underground railways. There were two strong arguments in favor of underground railways: legal—railways were prohibited in central London—and financial—the cost of property in central London was high and contractors would have to acquire and then demolish buildings to build surface railways. However, by laying underground tracks, contractors could develop routes that followed public roads and open spaces as much as possible to keep costs low.

Enter Charles Pearson, a lawyer and the solicitor to the Corporation of London. Pearson had already recognized the problems with traffic in London and had seen that it disproportionately affected the working classes, stating at a parliamentary commission on railway termini in 1846, "A poor man is chained to the spot. He has not leisure to walk and he has not money to ride to a distance from his work" (quoted in Kellett 2012, 34). In 1852 Pearson and others proposed a series of projects, including the construction of the Arcade railway, but the very nature of the project was resisted and described as "burrowing into the infernal regions and thereby disturbing the devil" (Dr. Cumings, cited in Halliday 2001). Ignoring criticisms, Pearson forged ahead with other proposals, finally receiving formal approval in 1858 for an underground train line that would become the

Metropolitan Railway. Approval was followed by a period of fundraising that resulted in a public-private partnership between the Corporation of London and two private railway companies (Martin 2012b; Wolmar 2004).

In spring 1860, workers began building the Metropolitan Railway using a cut-and-cover method.[6] The line included five stops along a three-and-a-half-mile stretch of the New Road (now Euston Road) in central London. Despite some months' delay and a series of complications, the line was inaugurated on January 9, 1863. Pearson, its great defender, had passed away four months before. One of his legacies was the implementation of cheap workmen's trains that made commuting financially accessible for the working class, or at least for those who lived along the lines that provided this service (Walford 1878).

The Metropolitan line quickly became popular in London. In its first day of service, it carried 30,000 passengers, and in the first six months it carried over 26,500 passengers per day. The media, originally scathing and skeptical of a project they deemed "utopian" and "an insult to common sense" started to perceive it as "the great engineering triumph of the day" (*The Times* 1861, 1863, quoted in Halliday 2001, 12).

Growth and Consolidation: The Subterranean Boom

The growth of an infrastructure is often, but not exclusively, achieved via competing systems. This was the case in London following the triumphant success of the Metropolitan line. Writing in 1878, fifteen years after the line's inauguration, Edward Walford noted, "There is scarcely any part of London or any of its outlying districts which cannot now be reached by rail, and by trains that are arriving and departing every few minutes."

Other lines sprouted quickly after the Metropolitan line, developed by competing rail companies. The expansion of the Metropolitan line was shortly followed by the construction of the Hammersmith and City line in 1864 and the District line in 1868; the Circle line was completed in 1884.[7] In the meantime, road improvements provided by the Metropolitan Board of Works in the 1870s and '80s somewhat reduced congestion, and the London General Omnibus Company continued to expand its profit margins with their omnibuses.

In those years, London's population continued to grow, expanding from 2.3 million in 1851 to over 7 million in 1911 (Wolmar 2004, 119). The working classes struggled to bear the economic burden of paying rent and commuting into work, except for in neighborhoods serviced by Pearson's workmen's trains. As the District and Metropolitan lines extended east and west, more affluent citizens moved further out. Meanwhile, employment rates increased in the West End, the City, and Whitehall, the latter of which had recently become the center of British civil service. London's streets were still struggling to keep up with the growing number of workers, businessmen, and tourists coming to the city.

At the time, the dependency on the cut-and-cover method for building became an issue. This method not only was labor-intensive and costly but also required closing down long sections of streets, further disrupting traffic in the British capital. The issue was bypassed with the invention of the Greathead tunneling shield, which enabled deeper tunnels and a new method of building underground lines. Surface traffic could continue on roads during underground railway construction, and deep tunnel lines began to be developed: the subterranean boom had begun. Around the same time, electrification (though originally slow to arrive in London) came to the Tube when the Waterloo and City line opened in 1890 and the Central line in 1900, whose bright lights made them serious competitors for the Metropolitan. Already forty years from its opening day, the Metropolitan was beginning to look dated.

Transport in London was also going through bureaucratic chaos as a new administrative body was founded: the London County Council, an elected local authority with a higher level of accountability to London residents. In 1891 the London County Council approved three projects for construction: an expansion of the Northern line and the creation of the Bakerloo and Piccadilly lines. The realization of these projects was dependent on a second prominent character in the history of London transport: a key managerial figure and financier named Charles Yerkes.

Often described as a man of few scruples, Yerkes dominated the underground railway world in London for a short but intense period. Yerkes was well known for raising vast amounts of funding for his transport projects through particularly complex financial maneuvers.[8] He raised £18 million to invest in building new underground lines and electrifying existing ones, a fund that is described by Martin as "an accidental Plan Marshall" (Martin 2012, 142). Yerkes's underground projects had been approved in 1891 and were well under way when he merged them under the company name Underground Electric Railways Company of London Ltd (UERL)—a significant step toward the forthcoming consolidation of the system and curbing competing networks. He expanded lines to service areas where he had property investments in London's suburbs. Consequently, the geographic borders between London and its suburbs became blurred, giving rise to the term Greater London, still used today.

Transport historians argue that there is much for which to thank Yerkes. Were it not for him, London might not have its dynamic Tube network, as, despite parliamentary approval for their construction, projects were delayed and often interrupted due to planning and financial difficulties (Wolmar 2004). Yerkes is also credited for initiating the unification of London's various underground lines, with the UERL being a starting point. Visually, the new UERL stations showed uniformity and branding, products of architect Leslie Green's design, with the exteriors in terra-cotta bricks and the platforms in white and green tiles.[9] Yerkes was also remarkably lucky with timing. The new lines were finished and running before the First World War broke out and halted railway planning and construction.

Lucky timing is a common theme in the history of transport in London. The development of the motor bus and the Tube were perfectly balanced such that one did not detract from the other: had the motor bus been a bigger success earlier on, the Tube might not have been developed as it did, and vice versa (Barker and Robbins 1975). In the 1890s, horse-drawn omnibuses had reached their peak, with over twelve thousand in service in 1895. Motor buses began to appear in the late nineteenth century, sparking the interest of the Motor Traction Company in 1899. However, this new technology only gained ground when petrol engines became a reliable technology and, importantly, when petrol became cheaper than oats for the horses. The London General Omnibus Company was quick to enter the motor bus game, afraid to lose their monopoly. Here, too, branding and uniformity began to emerge as they painted their buses red and made their vehicles identifiable with a logo of a spoked wheel with a crossbar and wings—an embryonic version of the well-known London Transport roundel.

In 1912, with Yerkes's passing, the new UERL director, Albert Stanley, acquired the London General Omnibus Company. This consolidation of two transport systems permitted the UERL to run buses in ways that benefited their own underground lines and created competition with those of other companies. Now controlling over half of the Underground in London, Stanley's UERL was in a good position to bring other operating companies together to discuss collaboration strategies. This would begin with a joint booking system, allowing passengers to transfer to other lines without purchasing a second ticket. This amalgamation of transport modes was called the Combine. Under Stanley's savvy press and publicity management, a new brand image appeared as blue plaques with "UndergrounD" in white block letters were installed at stations, both the initial *U* and the final *D* larger than the other letters. The first free transit maps recognizable to modern audiences, designed by Harry Beck, were produced by the millions. A prototype of today's roundel was developed: the circle and bar with stations' names on it. New signage was implemented to improve passenger flow, and in the 1920s and '30s, elevators at busy stations were replaced by escalators, which slowly became the norm for deep-level Tube stations (Pedroche 2013).[10] Having acquired the bus company, Stanley also extended motor bus routes to the suburbs, where they became brutal competition to the electric tramways that reigned in those areas but were subsequently closed.[11] Having secured control over much of the city's transit infrastructure, UERL weathered the First World War in a state of financial stability.

Momentum and Splintering: Administrative Changes

The rest of the twentieth century saw a slower pace of expansion while garnering increased ridership on the Tube system in London. The most significant changes were administrative in character. Below I have attempted to simplify from specific named organizations to considerations of public versus private administration of the network. While there were multiple administrative changes occurring to the system, daily passengers who were not familiar with the behind-the-scenes workings were largely unaffected except by fare changes over the years. Having reached this moment in an infrastructure's lifecycle, systems *appear* autonomous, even if they are anything but (Hughes 1983).

It became clear that the Underground was a "vital part of the infrastructure of the capital" (Wolmar 2004, 218), and the gradual increase in passenger numbers justified expansion proposals in the interwar period. As the UK Labour Party took the parliamentary majority in the House of Commons, the government passed the Trade Facilities Act of 1921 and then the Development (Loan Guarantees and Grants) Act of 1929 to stimulate infrastructure and large public works to create jobs. Both Northern and Piccadilly lines were extended, and the Combine continued to shift bus routes to integrate with the Underground, mapping onto rapid growth of the suburbs.

While daily passengers may have noticed the new roundels, the introduction of new buses, and, of course, the line extensions, the interwar period was more markedly characterized by administrative changes than technological ones, the most significant being London transport going from private to public ownership.[12] UERL had become a monopoly, fares were becoming arguably high for many passengers, and various routes were deemed useless. And while the London County

Council that had replaced the Metropolitan Board of Works in 1889 would have happily taken over administrative control from UERL, the lines had become too extensive outside of London for them be considered the right authority for the job. A new public corporation was developed in 1933, the London Passenger Transport Board (London Transport). London Transport was the amalgamation of the UERL and another "five railway companies . . . fourteen council-owned tramways, three private tram companies, sixty-six omnibus and coach companies and parts of sixty-nine others (Wolmar 2004, 266)," with a staff of over 70,000 people from drivers to mechanics to chassis manufacturers. London Transport, scholars argue, is "fondly remembered as the very apex of British public service" (Martin 2012b, 193), establishing itself as a truly recognizable consolidated system.

Administration of London Transport, however, remained unstable as the Second World War left the transport system significantly underfunded and in severe need of maintenance. Between 1933 and 2000, the management and oversight of London transport was tossed back and forth between the municipality of London and the British Department of Transport no fewer than four times, each time earning a new name. Under Margaret Thatcher's government, London Transport was taken over by the Department for Transport. This was done not to ensure public services, however, but rather to enable the privatization of transport operation and administration as new limited companies were created to contract out operations to third parties. Thus, the splintering that occurred not only was administrative but also marked the end of standardization in London's system. Until then, London Transport had owned its own garages and bus manufacturers, so all buses in London were standardized to their stipulations. The splintering resulted in the proliferation of bus models and a new approach to tendering contracts.

Finally, we reach the state of London transport as it is today, with one final administrative change. Administrative control was returned to the municipality of London in 2000, with the formation of the Greater London Authority and Transport for London (TfL), an important actor in our story in the next chapters. TfL is a complex bureaucratic monster with directorates and subsidiary companies as well as privately tendered contracts for various operations. Among more recent additions to the TfL roster of modes of transport is the Overground system: a complicated collection of railways fully orbiting London that has been renovating and modernizing since the early years of the twenty-first century. More recent still is the Elizabeth line, or what interviewees and I will refer to throughout this book as the Crossrail line. This has been a lengthy project in the making, deployed throughout the writing of this manuscript, and it will cut through London, east to west, connecting various suburbs to the heart of the capital.

USERS' EXPERIENCES

Though this account has been told and retold in various histories of transportation, it is interesting that the changes that have occurred to the infrastructure have not always been apparent to much of London's population. The cost of fares for accessing London transportation changed remarkably in the 1980s, which may have been the most obvious sign of the administrative shifts the system experienced. However, many of the shifts have been gradual and, to the public eye, many aspects of London transport have practically congealed: Harry Beck's map has endured (with addenda), as has the famous roundel and the vermillion buses. Public transport in London is, as one historian put it, "the language of the city, whether for Londoners or visitors" (Martin 2012, 270).

But is it the language of the city for *all*? It became clear in reading various volumes on the history of London transport (both popular histories and robust historical tomes) that disability was not a key consideration for the system builders or their chroniclers. Mentions of disabled people in those histories are few and far between, even in pieces written after the 1980s. The few mentions of disability by historians of transport are often laced with negative undertones, such as considerations of cost-effectiveness or unpleasant aesthetic changes. Garbutt, for example, discussed the new duty to provide accessible transport to disabled passengers in the 1980s as "a potentially difficult and costly obligation" (Garbutt 1985, 120). Martin, whose sensibilities are more aesthetic, is aggravated by audible announcements on the Tube system, calling communications "incessant" and due, at least in part, to disability discrimination legislation (Martin 2012, 216). Other books fail to mention disability at all, with glossaries and indexes lacking any mention of accessibility, disability, discrimination, step-free, or even the antiquated and discouraged term *handicap*.

On the other hand, to historians of technology, the pattern traced out here can be easily comprehended along the lines that Thomas P. Hughes (1987) draws in "The Evolution of Large Technical Systems." In this important text, Hughes identifies various milestones in the development of a system. These milestones are system building (through the invention of key system components and constituent parts that embed the invention into a niche, ultimately culminating in an innovative system), technology transfer and growth (where a large technical system spreads from place to place, acquiring diverse technological styles suitable to its locale of adoption and growing its number of users), to finally reach a period of consolidation and momentum (where competition between system alternatives is settled and the winner appears autonomous in their growth through various stakeholders' investment in their goals and directions). From the period of consolidation onward, other authors have picked up Hughes's pattern to add a moment of splintering of infrastructure from a single—often public—service provider to various private companies (Graham and Marvin 2002; Guy 1997). To Hughes, these phases comprise a "pattern of evolution" (1987, 56) wherein activities relating to one of the phases are more prominent in the network.

This is a useful pattern to be able to discern, and I was able to apply it neatly to the narratives offered in transport history books. It is useful because it tracks moments of expansion, enables the identification of system innovations, and traces the involvement of macropolitical actors (such as national and municipal governments) with a system. In essence, it outlines a compelling history of how the network has been pieced together over the years, from private enterprise to public consolidation, and the various in-betweens TfL eventually settled on in the past twenty years. For infrastructure studies scholars, this version of the story is often a useful one to lean on: when studying an infrastructure, knowing its development process and the key actors can offer clues to the values that have been embedded into the system. For example, tracing Margaret Thatcher's involvement in the shifting management of London transport in the 1980s clarifies why it went from a municipally controlled public corporation to an increasingly privatized one. Thatcher's government is already understood as one governed by neoliberal policies, so this emerges as an explanation for public divestment in transit. It is a first step to understanding the boundaries of the infrastructure one is setting out to investigate. For that reason, an experiential approach to studying infrastructures cannot do away entirely with this classic approach. Thus, the historical double bind of infrastructure comes to the forefront—while classic histories do not contend with the experiences of users, users of these infrastructures have to contend with infrastructural histories.

A crip feminist lens on transportation in London therefore requires us to engage with the classic story more critically to ask, How are echoes of the past shaping the experiences of users? What other histories have influenced the shape of the infrastructure that might not otherwise be accounted for in classic histories? In so doing, we can begin to account for what interlocutors have identified as the Victorianness of the system that the infrastructure has inherited over the centuries.

A Critical History of Transport in London

If the focus of this book were transportation writ large, with a particular focus on a functionalist narrative of mobility in the British capital, the narrative offered above might constitute a satisfactory approach. The tale of an infrastructure's historical development is often told from the standpoint of system builders, and while more anthropological approaches have undoubtedly assisted in teasing out the tensions that exist in these processes of development (and subsequent exclusions), there are no extant works dedicated to understanding the history of transportation in London in relation to wheelchair users' access needs, nor to disabled persons more broadly. A first step to redress this, therefore, is to offer an account that is centered on disabled people in relation to transportation. To do so, it is useful to recover the history of disability in the context of the United Kingdom, giving particular focus to the Victorian era and to moments where historical evidence enables the tracing of the two histories. As I lay out some of the history of disability and disability activism, I will focus on how this history has overlapped with transportation history, showing how it is through disability rights activism that conceptions of accessibility in public transportation have shifted as advocates have successfully widened the claims to access. Finally, I will argue that the widening scope has also been limited through the infrastructure's historical inheritance and stabilization.

The Deserving Poor: Access as Productive Agency

The history of disabled people is a patchwork quilt being recovered and mended by many thoughtful scholars who have made it their life's work to contextualize the spaces that disabled people have occupied and been made to occupy over the centuries. Colin Barnes and Geoff Mercer, in particular, point to how the history of disabled people has often been rewritten from "an overtly individualistic medical perspective" (Barnes 1997). It was with the rise of disability studies in the 1970s and 1980s that this history began to be reclaimed in the United Kingdom.[13] Authors such as Vic Finkelstein and Colin Barnes provide us with a solid starting point to help parse out more clearly what Carl alluded to in our interview—that "obviously" disabled people weren't considered important in Victorian times.

British disability history is still very much in the process of being constituted, and the gaps in archives are significant. However, historians thus far have emphasized, in the UK context, the construction of disability as an object of charity, making disabled persons the vehicle of others' goodwill rather than being perceived as fully realized individuals. The development of a market-based capitalist economy in thirteenth-century England established this "charitable status" through the Statute of Cambridge in 1388. This statute was a result of labor shortages after the Black Death, when wages for laborers increased and workers started fleeing their landowners to become freemen. Significantly, it set the scene for subsequent legislation that created a distinction between "sturdy beggars," those capable of working, and "impotent beggars," the elderly and those with impairments. This served as a basis of definition for the "deserving poor"—those who would be looked after by charity coming primarily from monasteries and churches. These definitions would endure through subsequent Poor Laws.[14] The better-known of these was the Elizabethan Poor Law, officially titled the Act for the Relief of the Poor of 1601. This act placed responsibility of caring for the impotent poor (defined as "the infirm, the elderly, and children") on the state, at that point organized at a local parish level. The lives of disabled people became governed by state administration, and disabled people were depicted as indigent and passive.

The mid-eighteenth century broadly marks the beginning of the Industrial Revolution, characterized by increased and sustained material and industrial production, as well as the breakdown of the rural state and church welfare in the United Kingdom. The changes in London's population were briefly described in the beginning of this chapter, but it bears repeating that cities experienced a significant population growth in the nineteenth century, with many rural migrants moving to live closer to where factories were being built.

By the late eighteenth century, the size of machinery had grown due to the drive for bigger and more efficient production. It was also the period of the rise of coal and mining exploits, as well as new and more efficient farming methods, such as the threshing machine. A crucial consequence of these shifts in the mode of production was that the "average worker" was created, a concept referring to physically able persons who could operate these new technologies, excluding disabled people not only as indigent but as having nonstandard, nonproductive, and nonnormative bodies. This argument is supported by many disability studies scholars in both British and American traditions. We see the emergence of the nonstandard or disabled body as being the literal misfit of bodily capability and function with production demands, creating compulsory able-bodiedness (McRuer 2010) and subsequent understandings of a neoliberal ableist society

(Goodley 2014). Given the nature of the new machines, home labor or artisanal work became a less viable option. Laborers had to displace themselves to their place of work. This, Finkelstein argues, was the "decisive push which removed crippled people from the social intercourse and transformed them into disabled people" (Finkelstein 1981).[15] One's social status became defined by one's relationship to the mode of production, to machines, and disabled people literally did not fit this paradigm of production. They began to be perceived as unproductive members of society.

The perception of disabled persons as unproductive gained ground throughout the Victorian era. Several tendencies reinforced that perception in that era. For example, there was renewed intensification of previous legislative practices described above. British Parliament passed what became popularly known as the New Poor Law, or the Poor Law Amendment Act, in 1834. This new version of the Poor Law sought to centralize the parish-level system of poor relief through the construction of workhouses. Purposefully designed so that conditions within the workhouses were worse than the conditions outside of them, these spaces were meant to attract only people who truly felt that they had no other way to survive. A further condition of entering the workhouse was the workhouse test, through which persons were evaluated as being capable or incapable of work. If deemed capable of sturdy labor, a person would integrate into the workhouse system, essentially working for their keep. Here, again, the litmus test of capacity for labor was being applied, with disabled persons often deemed as falling short of that capacity.

Thus, disabled people were often seen as requiring protection and charity outside the scope of labor and were either placed in institutions or limited to their homes. In many cases, families or charities were the primary recourse to enable some form of independent living. Indeed, disabled people were often removed from their communities entirely, perceived as burdens to their caretakers, and placed in increasingly specialized institutions. Through this process they were no longer solely subjects of charity; they also became objects of study.

The Victorian era also marks the advent of the professionalization of medical occupations and, with it, the medicalization of the body and mind. Authority of biological and medical knowledge resulted in social constructions of the good, healthy, and normal body as bodies themselves became evaluative objects (Barnes 2011; Snyder and Mitchell 2001). In 1829 Francis Bisset Hawkins began to apply statistics to medicine; across the English Channel, Adolphe Quetelet began to apply statistics to measurements of the human body. Through Quetelet's development of the concept of *l'homme moyen*, the average man, and the publication of Charles Darwin's *On the Origins of Species* in 1859, Francis Galton found the building blocks of what he would name *eugenics*. Eugenics had its talons in British intellectual circles, including in some often perceived as progressive. John Maynard Keynes, for example, famous for his opposition to laissez-faire economic theories and politics, was the director of the British Eugenics Society between 1937 and 1944. In giving the Galton lecture in 1946, Keynes noted that he believed nothing mattered more to the human race than "the possession of a sound genetic endowment" (Keynes 1946). Winston Churchill was also an enthusiast, fearing the influence of the "feeble minded" on future generations (Woodhouse 1982). This influenced British government policies as the notion that society should care for disabled people began to be seen as an economic burden, an accounting problem to be solved. The distinction between an *us*, a nondisabled productive class, and a *them*, the deserving disabled poor, continued to grow.

It is important to be explicit that the Victorian era is the era of the birth of the eugenics movement—a movement that would naturalize and underline the need for a systematic exclusion and sterilization of disabled persons for the alleged betterment of the human species. Hierarchical notions of human capacity, behavior, and aesthetics found scientific support, with an accompanying fear that disability as an inherited characteristic might weaken humanity. Though this movement gained a name and widespread support in the Victorian era, it did not mark the beginning of something novel or particularly radical. Indeed, eugenics is rooted in practices and beliefs that much predate it, as shown above in the histories of "deserving" and "undeserving" poor, and the accompanying segregation of persons perceived as unproductive or incapable. But it was in an era concerned with the superiority of some bodies over others and with the continual growth of industry and productive capacities that London's transportation system was conceived.

The birth of London's transport system is, therefore, already historically entangled with conceptions of access. *Access* here, however, is primarily concerned with access to labor opportunities, access to London as the epicenter of British power, and access to leisure. The developers of the Metropolitan line were focused on very particular mobilities around the capital, and mentions of disability, or access of disabled passengers, in the classic narrative of the birth of the Tube are nowhere to be found. Theirs is not the accessibility that the Metropolitan Board of Works or British Parliament were concerned about; transport administrators' primary focus was on easing gridlock to facilitate the flow of a particular conception of prospective travelers to key termini around London. Pearson, who shared these concerns, had an additional worry: the question of access to jobs. The legacy Pearson left to the transport system with his workingmen's trains gives a clear sense of whose access he wanted to ensure—access of the working class to the new industries of the Victorian era. Pearson arguably understood access as a financial category that could be remediated through discounted rates. The birth of passenger transport in London was therefore related to one that underlined the importance of flow of services and goods throughout the capital, and of the persons related to that specific provision of those services and goods.

To change infrastructural access for disabled passengers, the system needed a different conception of passengers altogether. Who is welcomed, perceived as persons who should be able to move freely throughout the capital? Who are members of the "public" that uses "public transport"? Victorian ghosts, and the eugenic legacy that would last well into the twentieth century, thus excluded disabled persons through a relatively simple move: by not perceiving them as the public at all. It was not even a question of financial access that might be rectified through the provision of cheaper trains; the system did not even consider the possibility of their needs, as they were not imagined users.

It is only by centering the history of disability and the disability rights movement and its interrelation with the history of transportation that we begin to unearth how dependent the latter is on the former. In particular, the history of transportation in London could have been significantly different from the 1970s onward if the disabled rights movement in the United Kingdom had not pushed to redefine conceptions of *public* and *access*.

Access through Segregation

The shift in conceptions of accessibility that created new forms of access to transportation for disabled persons was dependent on shifting perceptions of disability tout court. The two world wars provided the first gentle pushes to change public perceptions of disabled people, with an exceptionalism afforded to disabled veterans who returned home with physical and mental impairments (Woods 2005). The British government began to see itself as responsible for the wounded and took a more active role in ensuring the employment of veterans. The Ministry of Pensions was set up in 1916 and became responsible for not only veterans' pensions but also their medical rehabilitation (Cohen 2001). However, this was done according to highly medicalized and standardized definitions of the body with the goal of normalizing the body once more. Wheelchairs were, ideally, to be momentary props. If permanent, they were perceived as a failure of the medical process that had not offered a dearly sought cure to an impairment. Special employment facilities were organized, largely under the Disabled Persons Employment Act 1944, and sheltered workshops were built. The legislation also codified preferential treatment of former servicemen and servicewomen, as well as the definition of two types of disabled people: those who were suitable for mainstream employment and those who were not. Wages at segregated workhouses were low, £90 weekly, compared to the average nondisabled laborer's salary of £200 (Barnes 1991).

In the mid-twentieth century collective disabled voices demanded change. Residential homes for disabled people provided an alternative to institutional care or being left at home as new disability charities began to be established during the two decades following the second war. Unlike previous organizations such as the Royal National Institute for the Blind (set up in 1868), these new charities were founded by disabled people and their family members.

Direct action would become important in the 1960s as demands for appropriate employment and housing intensified. In 1965 the Disablement Income Group (DIG) was established, a pressure group that contacted sympathetic members of Parliament and peers. DIG organized a series of disability rallies in Trafalgar Square between 1966 and 1968. This was the era of civil rights movements in the Western world, with identity politics coming to the forefront. Disabled people also took to the streets and self-organized, becoming one of the new social movements of the late twentieth century, if one of the ones less often discussed in the United Kingdom. It is these movements that begin to excise some Victorian ghosts.

Thanks to the work done for and by wounded veterans, including state aid to impaired veterans following public demonstrations, larger numbers of disabled people felt enabled to step out and speak to their personal experiences in the 1960s and '70s. This coincided with the genesis of disability studies as an academic field in the United Kingdom, the first wave of which consisted of work by Vic Finkelstein, Michael Oliver, Paul Hunt, and many others. Hunt had published *Stigma: The Experience of Disability* in 1966, and was at the time working toward shifting definitions of disability from ones mired in medical assumptions about the body to an approach that centered disability as a socially constructed and historically oppressed category. In 1972 the *Guardian* published a letter in which he called on disabled people to form "a consumer group" to consolidate their movement and advance their demands. In response, the Union for the Physically Impaired Against Segregation (UPIAS) was formed and a new definition of disability was coined. This gave birth to the social model of disability in the United Kingdom through the influence of the independent living movement and the Disabled Students' Program in Berkeley, California, that established the first Center for Independent Living. These centers, run by disabled people for disabled people, emphasize independence and civil rights. They provide services to empower and enable disabled people, facilitating their integration. Services include the provision of information, assistive devices, and contacts with work opportunities

In 1970 the Chronically Sick and Disabled Persons Bill became an Act of Parliament. The act required that local authorities provide aid and welfare to disabled people in the form of access to recreational activities and public buildings, among other stipulations. This act, which passed two years after the construction of the Victoria line, did not include clauses on access to public transport. We thus see a continuation of the pattern through which mobilization to improve the quality of life of disabled people was done through segregated environments—separate workhouses, separate schools. This included segregated transportation, with the introduction of some door-to-door transport options like London's Dial-a-Ride services in 1982. With the advent of the private car and some adaptability in its technology, the problem was further individualized as disabled people became enabled to drive themselves or be driven by caregivers; they are, today, still largely dependent on private transport (Barnes and Mercer 2010). Public transport was rarely perceived as an option.

Yet in using the social model of disability, the disability rights movements became stronger in the 1970s and '80s. As Beckett and Campbell argue, though social movements are often heterogeneous, having a common goal to march toward allows for cohesion in overall demands. The social model functioned as an oppositional device that "allow[ed] for the refusal of the forces of subjection" (Beckett and Campbell 2015). In this way, disability rights movements identified a common enemy: a disabling society. The social model of disability became a frame of reference and "basis for political mobilization" (Blume and Hiddinga 2010, 228). What makes a person disabled is not their impairment but the circumstances surrounding them. This was a powerful shift in rhetoric that made possible demands for concrete changes to the environment. The authors of the second wave of disability studies, such as Tom Shakespeare, quickly worked toward equating and integrating disability with other social rights battles, such as those of race and gender, though it ought to be noted that, within disability studies scholarship, a significant amount of debate has also been afforded to dissecting the uses and limits of the social model.[16]

It was under Margaret Thatcher's regime that the first disabled persons' stakeholders group relating to transport was born. The Transport Act of 1985, which simultaneously abolished the Greater London Council and brought London transport under the remit of the national Ministry of Transport, also created the Disabled Persons Transport Advisory Committee (DPTAC). This committee would be responsible for providing an annual report, advising

> as to measures that may be taken with a view to—
>
> (a) making access to vehicles used in the provision of public passenger transport services by road easier for disabled persons; and
> (b) making such vehicles better adapted to the needs of disabled persons
>
> *Transport Act of 1985*

Indeed, it was through DPTAC's advising that the Docklands Light Railway was built as London's first mode of transport to offer level access boarding, where the train floor and platform are at equivalent heights and with a minimal gap, more easily enabling boarding and alighting. This opened the door to a shift in understanding of whom accessibility was for. Until the constitution of this body, access had been offered to disabled persons largely through the provision of segregated environments, from veterans' workhouses to schools to transportation. Though the kernels that enabled questioning this approach were already present in much of the independent living movement, in the world of transportation in the United Kingdom, it would require activist mobilization to expand them further.

Equality in Access

The creation of DPTAC provided the impetus for disability rights activists to expand their demands to include integrated modes of transportation—in other words, modes of transport that were not of a segregated nature. As Thatcher's administration gave way to Tony Blair's New Labour government, bold new disability organizations were born, with accompanying new strategies. In 1992 Disability Awareness in Action (DAIA) was established, and the Disabled Persons' Disability Action Network (DAN) followed in 1993. The prominence of the word *action* in both titles is a telltale sign of these groups' purposes and aims. DAIA and DAN staged over a hundred demonstrations demanding "rights, not charity." Using civil disobedience and nonviolent disruptive methods, DAN blocked Abingdon Street, across from Westminster, in 1995 and threw red paint on the steps of Downing Street in 1997, symbolizing the blood that would be shed by disabled people should the British government go through with its proposed benefit cuts (Oliver and Barnes 2006). While many organizations for disability rights in the United Kingdom had, until then, largely focused on questions of financial access (i.e., securing access to disability assistance from the government), these new organizations expanded their demands to encompass access to a wide range of built environments as well as financial inclusion.

It is worth remembering that the beloved red double-decker buses, some of which were still Routemasters, remained inaccessible to wheelchair users in the 1990s, with at least one step at all entrances to the lifted chassis. DAN thus spearheaded the first campaign in the United Kingdom that focused exclusively on accessibility to public transportation. Between 1990 and 1993, DAN led the Campaign for Accessible Transport (CAT), with demonstrators chaining themselves to buses at key London locations, causing traffic jams throughout the capital. Perhaps the most emblematic sign and slogan of the campaign was a riff on the Star Trek opening scene: a black cloth decorated with a starry sky, planet, and bold pink and white words reading "Disability rights, the final frontier. To boldly go where all others have gone before. Disabled People's Direction Action Network, Rights Now." We can identify here a tongue-in-cheek understanding of access as that which enables one to go places where other, nondisabled people have already been. Access is thus constituted as something that is specifically out of reach for this population—disabled persons. Between pressure from DAN, lobbying by various disability organizations, and DPTAC consultations, inaccessible buses were progressively phased out of London's transport system beginning in 1994, as London Transport trialed routes with low-floor buses.

The Disability Discrimination Act (DDA) was passed in 1995. A series of new "firsts" for disability access would emerge in the late 1990s, showing that the term *accessibility* was beginning to encompass broader understandings of *access* and *public* in London. Changes included a significant extension of the Jubilee line, new trains that enabled some level access at specific stations, color contrasts on signage for blind and visually impaired passengers, and dedicated wheelchair areas. These changes occurred largely under the remit of Transport for London, which is outlasting the average lifetime of all of its twentieth-century predecessors![17] It was under Transport for London that the last Routemaster was finally taken out of service in December 2005—with some nostalgia from proponents, but prominent signs saying "Good Riddance!" waved by disability rights protesters (Associated Press 2005). The same year, the DDA was finally expanded to include protection against discrimination on land transport, and its definition of *disabled person* shifted toward one more informed by the social model of disability.

The disability rights movement was thus actively involved in shaping transportation in London. The influence of stakeholder groups such as DAIA and DAN was brought to bear on various modes of transit in the city, particularly in their articulation of access not as something that is easily achieved through "equal but separate" solutions but rather as something that ought to be integrated into that which is already "public." This shift in rhetoric and demands disrupted how the infrastructure understood its own constitution of accessibility, spurring significant changes to provisions of what it now newly understands as *access* over the decade that follows.

The Limits of Reasonableness

By 2010 the United Kingdom had become a signatory of the United Nations Convention on the Rights of Persons with Disabilities and the DDA was absorbed into a single piece of legislation: the Equality Act of 2010. This legislation was the result of merging previous acts such as the DDA, the Sex Discrimination Act 1975, and the Race Relations Act 1976. The Equality Act defines age, sex, gender, disability, religion, and sexual orientation as "protected characteristics." Of these, disability is the only one to have a dedicated "adjustments" section, wherein it is stipulated that, if a disabled person is faced with significant disadvantages, "reasonable adjustments" must be made by the person(s) deemed responsible for the presence of those disadvantages. We will return to the question of reasonableness soon.

In direct response to the changing landscape of disability rights and legislation, various changes were made to London's transportation to accommodate disabled passengers explicitly. To point out these changes is to do justice to the narrative of wheelchair users interviewed who were happy to recognize these shifts as positive indicators that perhaps the infrastructure is moving toward better inclusion. Indeed, in the 2010s, British Parliament held a consultation on "access to transport for disabled people" that highlighted the importance of accessible transport as a means of inclusion for disabled people. It was clear in this report that one of the key moments of change for understandings of accessibility in London was the 2012 Olympic and Paralympic Games. Such were the Games' perceived importance that one of the key questions of the report was, "What can be learnt from transport provision during the Paralympics and how can we build on its success?" (Transport Committee 2013, 5). The London Organizing Committee of the Olympic and Paralympic Games was praised by various organizations and individuals, and early in the planning process for the Games, committee members placed transport and accessibility as priorities on their agenda (LOGOC 2008).

Between 2007 and 2012, various improvements in accessibility were brought to London's system, including the overhaul of the London Overground in the east of the capital (primarily where the Games were hosted). Multiple stations were refurbished to become step-free ahead of schedule, and new signage, lighting, and induction loops were introduced at 175 stations. Of specific interest to wheelchair users was the installation of raised humps on platforms to ensure access from platforms to trains and the introduction of manual boarding ramps at various stations. These ramps would be staff-deployed and required booking 24 hours in advance to ensure an employee would be present for their use.

These efforts did not go unnoticed. Many wheelchair users interviewed pointed to the 2012 Games as a key moment of change. As Sophie told me:

> Well, I've started to use [public transport] a little bit more since the Paralympics, mainly because, one, I wanted to go to the Games, and two, there was such a big deal made about the support and the accessibility of it all, etc.

The public-facing document titled "Your Accessible Transport Network" included a foreword by then-mayor of London Boris Johnson, presenting the Games as "the most accessible ever held—with more disabled people travelling to more events at more venues and locations than on any previous occasion" (Transport for London 2012).[18] It was at least in part on the back of this "big deal" and the successes of the Games that further accessibility efforts gained a foothold and disabled activists leveraged additional demands from TfL. We see new accessibility efforts rolled out more permanently in London from 2012 onward. Between 2012 and 2015, the provision of manual boarding ramps was extended from sixteen stations to fifty-five.[19] TfL also finally did away with requiring advance booking for the manual boarding ramps service, replaced with "turn up and go" services—arguably also a victory following from the campaigning of disability activist groups. More stations are being made step-free yearly, and new low-floor S-stock trains on lines that serve significant central London stations such as Paddington and King's Cross stations (the District and Neapolitan lines) have further enabled this form of access.[20]

The newest project discussed above, the Crossrail (now known as the Elizabeth line), became a specific hot spot of disability activism in London in the 2010s as it was originally going to retain seven inaccessible stations. As will be discussed in more detail in subsequent chapters, intense mobilization by disability activists succeeded in pushing all stations to be made wheelchair accessible. Whether renewing old stations requires compliance with accessibility regulations has been a point of significant contention, but, slowly, the number of step-free accessible stations has reached 92 on the Tube network (one-third of the 270 stations) and around 60 on the Overground system (about half of the 112 stations) at the time of writing.[21]

Thus, prior to the worldwide pandemic of 2020, TfL made a wide range of changes to London transport's accessibility, and many wheelchair users recognize these changes:

> But I think the differences I've seen in London since I've moved here has been tremendous. This place that we're here now, King's Cross St Pancras, I use this station all the time and it's brilliant. —*Faith*

> So the whole of my narrative includes a sense that things are getting better; Tube stations, Tottenham Court Road is going to be fabulous. There's a couple more in the pipeline. —*Anton*

Yet despite these shifts, wheelchair users' descriptions and experiences of the transport network remain contradictory. Or, as Char Aznable said, "TfL services are improving, but spotty." Thus, while the improvements are recognized, this recognition is often qualified by discussion of the barriers that continue to exist. Perhaps particularly interesting are the ways in which even so-called improvements to the built environment become tensions or problems faced by wheelchair users.

Here we begin to encounter the limits of the shifts in understanding of accessibility: despite the work of the past two decades, and despite improvements that wheelchair users themselves are quick to recognize over the years, barriers to access remain. The present has been made to contend with its Victorian past through the work of making changes and shifting the infrastructure's understanding of access, but it has not sufficed to integrate access into the system seamlessly. As Bess Williamson articulates in *Accessible America*, at least part of the difficulty in these shifts can be traced back to how even disability activists' slogans and demands reinforced the idea of a "neutral" spatial politics, as if everyone else had already been included in spaces not previously integrated into the system (Williamson 2019).

The wording of legislation, in some regards, constricts changes to the infrastructure. Both the DDA and the Equality Act that supplanted it included a reasonable accommodation clause.[22] The Equality Act states that when a disabled person is placed "at a substantial disadvantage in relation to a relevant matter in comparison with persons who are not disabled" that the responsible party has "to take such steps *as it is reasonable* to have to take to avoid the disadvantage" (emphasis mine). In essence, the very idea of a reasonable accommodation is one that limits accommodations to what is practicable within constraints—or, at the very least, offers itself up to legal interpretations that require an additional layer of navigation. It becomes constrained to what, within its context, is deemed feasible when considering a variety of criteria: the size of the organization responsible, the amount of money it would cost, what changes have already been made for accommodations, how many people would benefit from the change, and more. In the context of public transportation, various constraints can be argued to already be in place for the implementation of change. Among them is simply the age of the system itself and how it has consolidated over time.

Changes to the infrastructure for disabled people are measured against the infrastructure itself. For example, due to the cut-and-cover method used for many of the older lines of the Underground, various stations have somewhat curved tracks that follow the lines of the streets over them. The straight train wagons on these curved tracks are, at least in part, what create the well-known gap in some London stations, impeding access for those who have difficulties stepping over the gap. To amend this accessibility issue by, say, closing the stations to straighten the tracks and platforms so the gap is minimized might be considered an *unreasonable* adjustment: it would demand tremendous amounts of money and significant line closures throughout the system and would affect thousands of journeys over months of disruptions. Other solutions might therefore be proposed as more reasonable alternatives: the provision of manual boarding ramps at stations that already offer access to the platforms, for example, so that access can be made to the trains themselves.

It is through this teetering balance of *reasonableness* that accessibility must be negotiated, often using retrofits to the system. I define *retrofits* in this book as the attempt to shore up two histories in the present and bring an infrastructure fully into its new contingencies. Temporality lies deep in the term *retrofit*, which alludes to fitting something of the past with something new to keep it working. Importantly, *retrofitting* is not the same act as *replacing*, which, as discussed above, would be impossible in the lives of infrastructures. It is a deliberate move to tweak the infrastructure, hack it, improve it enough to keep it alive, functional. And whereas Howe and coauthors focused primarily on the future-facing orientation of the act of retrofitting, as "look[ing] to past projects—failed or successful—to foresee what comes next" (Howe et al. 2016, 555), I see in it more of an eye to the limitations inherited from the past that ought to be fixed to enable survival into the future. In that regard, a retrofit is a temporal as well as a material moment that demands specific interventions in a network. It is this material context of retrofits, and of an infrastructure more generally, that we need to address next.

TWO / Designed Materiality / *"Stop sending me apologies. I want change."*

It was in 2005 that the last of the Routemasters left service in London, leaving in its wake a fleet of low-floor buses with wheelchair-accessible areas. I met Anton ten years after that, on a cool August afternoon. A white man in his late fifties, he joined me at one of University College London's wheelchair-accessible buildings, using a power wheelchair. We settled into our seats, and he pulled out notes he'd prepared. Anton was the only one of my interviewees to have prepared, prior to the interview, a list of things he wanted to speak to me about.

> I can just go through my notes and talk to you? So, my experience with public transport in London? Okay, here we go.

He began with the Tube. Most of zone 1, central London, he deemed useless. "The Tube is more or less out of bounds," he continued. The elevators are unreliable, and the labels indicating wheelchair accessibility on the maps were inconsistent in his experience. He continued down his list. The Overground: not bad. Pavements, suburban trains, people in general . . . Then, with a tone of finality:

> So now . . . my pet hate. Buses. You must have heard a lot about buses.

We laughed as I said that I had, indeed, heard quite a bit about buses.

> You're going to hear it all again now.

Anton was the thirty-first person I interviewed for this work, and much of what he told me corroborated, supported, and expanded on what other wheelchair users had already told me about their experiences with buses. His ordered approach highlighted issues he has faced with bus drivers, with other passengers on the bus who refuse to make space, with the design of older ramps on buses, and with the limitations of the Equality Act, and how all of this, ultimately, boils down to a refusal to spend money on disabled persons' needs.

> So that's me rambling about law and money, a bit tangential but it's . . . it's also the heart of why traveling around London is difficult and why there's no redress, really, in practice. Complaining to TfL is one of my full-time occupations. Not really. It's a part-time hobby. I am so sick of getting back from TfL apologies and bland assurances. I should have brought some around. I'll read you one from my email. It's . . . it's just . . . At one stage I said to them, "Stop sending me apologies. I want change."

Anton's frustration and desire for change seem at odds with the histories that I shared in the previous chapter. I argued that transportation in London has been molded by the work of disabled passengers and activists over the past thirty years. I pointed to some of the retrofits of the system. Anton, on the other hand, is saying that he wants change. The goal of this chapter is to focus on this seeming paradox: how is it that so much change seems to have occurred, and yet passengers who use wheelchairs still experience frustration?

While the work of disabled passengers and disability rights activists of the 1990s has enacted change in conceptions of accessibility in London's transportation infrastructure, I hinted in the conclusion to chapter 1 that the work of retrofitting is limited in its impacts. Simply put, retrofits are an infrastructure's attempts to bring itself into alignment with present contingencies and demands. I argue that retrofits are not attempts to see into the future but rather attempts to make parts of the past more relevant or appropriate to the present. In this chapter, I further argue that the relative plasticity of infrastructures is limited by what I define as an infrastructure's designed materiality. The designed materiality of infrastructures is the compounded material realities of the rigidity of the materials used for building the infrastructure (cement, iron, glass, etc.) with what has become naturalized about their particular shapes and design over time—the knowledge materialized and petrified by the materials themselves. Wheelchair-using passengers already begin to identify, in their own discourses, evidence of the resistance that infrastructures have to change:

> And I think that we are reluctant to change, we were in this . . . in Britain, because we like things, we like traditions, we like things the way they are.
> —*Faith*

> There's a lot of history in Transport for London, there's a lot of cost when put in practice, there's a lot of, "Well, we do it this way because we've always

done it this way and we can't change it." And that's often the hardest thing you've got to do, you've got to break down. —*Alan*

These are striking terms, of feeling that one of the biggest barriers to enhanced accessibility is the need to bypass and "break down" sociocultural traditions. It is a clear illustration of how materiality is often represented as rigid and unforgiving when, once we dig into an infrastructure's history, it could always have been otherwise. We saw how that was true in chapter 1. There, we were tracing ghosts, where they emerged and where they remain. In this chapter, we contend with the material ways in which these ghosts remain, where they have sunk deeply into the infrastructure, and how current wheelchair passengers tackle them and, in so doing, change the infrastructure gradually.

When one interacts with an infrastructure, one is interacting with its *designed materiality*—an environment that is not neutral, that is shaped by its social history as well as the physical properties of the materials that constitute it. I argue that the historical work of shifting conceptions of accessibility done by disabled activists as part of this infrastructure's social history have affected the system's designed materiality through the introduction of retrofits. Retrofitting is necessarily limited in the context of infrastructures, largely because retrofits address only one aspect of designed materiality, such as easily targeted access icons, like ramps, without due consideration to how designed materiality more broadly shapes and mediates actions and behaviors beyond those artifacts. I use the metaphor of niche construction to highlight how wheelchair passengers are often making material arrangements otherwise for themselves through what I call *belligerent techne*—that is, a particular formulation of knowledge and practices that go against the inscriptions of a system so as to weave forms of accessibility into being on their own terms. Despite these interventions, their work is constrained by the infrastructure—the environment they are sunk into. I then offer the metaphor of niche construction, borrowed from evolutionary biology, to describe how belligerent techne shape infrastructures. In using this metaphor, I hope to provide an alternative path to understanding the scalar concerns at hand, showing how agents effect change in their infrastructural environment. This enables a subversion of both top-down and bottom-up theories of change, trumping the agency versus structure debates so commonly found in sociology. Niche construction as a metaphor thus becomes a path toward thinking of infrastructural development and reconstruction. Niche-constructing work enables me to highlight, ultimately, how infrastructuring work is knowledge work in chapter 3.

SOCIOHISTORICAL MATERIALITY AS DESIGNED MATERIALITY

Scholars have long attended to the material dimensions of infrastructure because they rightly acknowledge that the specific properties of materials enable and constrain a given infrastructure's mode of existence. To think of "the physico-chemical dynamics of specific materials, including their propensity to corrode, crumble and fracture" (Barry 2020, 99) or ask what "the tensile properties of iron permit" (Anand, Gupta, and Appel 2018, 9) is thus integral to critically reflecting on infrastructures. But these materials—cement, iron, glass, rubber, and more—should not be treated as neutral; they have their own lifecycles and are entangled with their own social biographies of aesthetics, promises, and functionality. Penny Harvey tells us that materials "display a certain autonomy" from these social, economic, and political concerns (Harvey 2015), which is a significant aspect to consider in the temporal and historical timelines of infrastructures—their maintenance, repair, decay, and even death. Nevertheless, a focus on these autonomous forces, no matter the importance they might bring, without due attention to the form of these materials risks naturalizing an infrastructure's design in a way that occludes its social mediation and origin. We must always bear in mind that infrastructures are artifactual systems with particular forms that in turn constrain, enable, and afford our experiences with and access to them. A study of the vitality or vibrancy of materials alone, as vitalism-focused new materialism has made possible, misses a key humanist concern: namely, that the promises of matter and their formations—thus, the promises of infrastructure—are not distributed generically or equally. When Harvey suggests we trace "what economic and political trajectories" materials congeal, I see her implicitly urging us to consider also the forms into which materials are congealed and to tend to the consequences of those forms. Studying the particular shapes these materials are made into and the ways they have been designed as technologies has mostly remained in the remit of the sociology and history of technologies and artifacts and rarely emerges as an explicit concern in the field of infrastructure studies. Here, I propose to extend that concern to the material design of infrastructures themselves.

Science and technology studies have always been concerned with issues of design. A quintessential example is Langdon Winner's article "Do Artifacts Have Politics?" in which he asserts that we ought to concern ourselves with "the characteristics of technical objects and the meaning of those characteristics"

(Winner 1980, 123).[1] Both of the popularly used examples—Robert Moses's bridges and the University of California's mechanical tomato harvester—are ultimately concerned with what each technology enables or constrains. The salient point of Winner's essay is that the bridge had an explicit design aim of exclusion whereas the tomato harvester case ultimately favored large private agricultural companies due to the size and cost of the machines. Of course, as Winner points out, one might quickly ask questions related to intentionality: Moses *intended* bridges to stop buses (and, therefore, lower-income Black residents) from reaching the beaches, but there was no intentional plot in the design of the tomato harvester.[2] This intentionality is ultimately of minimal interest to Winner and is also largely irrelevant to our case. The reality of design is that these "characteristics of technical objects" result in orderings of choices (more or less intentional) in forms, function, and aesthetics. STS tells us that different arrangements of forms and functions further result in different configurations, or scripting, of users (Woolgar 1991; Akrich and Latour 1992). A fence in one location, for example, renders passage beyond it difficult. Thus, diverse material arrangements and configurations translate into differential orderings that, in themselves, encode different orderings of power and oppression (and, to make the classic STS move, are themselves the result of orderings of power).

I use the concept of designed materiality to capture these differentially entangled histories and orderings of the materiality of infrastructures while contending with how the form and shape of these infrastructures are the conduits through which multiplicity occurs and creates different experiences and diverse relationships. It borrows from both feminist new materialisms and American pragmatism to acknowledge that an infrastructure's material characteristics not only are related to questions of the perdurance of materials but also are filled with meaning; they are literally meaning*ful*. Designed materiality thus forces us to think more carefully about the ways any infrastructure is historically and socially inscribed. London's transportation system is historically and socially informed; its material shape has been produced by social values, prejudices, and cultural assumptions. In this sense, as feminist materialism reminds us, matter is never flat or singular. Matter is multiple. In Barad's words, materiality is "a dynamic and shifting entanglement of relations, rather than . . . a property of things" (Barad 2007, 35). I would extended Barad's argument to say that materiality is *both* the shifting entanglements of relations *and* the property of things. The properties of things, as any designer or engineer knows, contain in themselves choices pertaining to both form and material. These choices take shape in relation to an imagined or postulated user, and these shapes afford or enable particular actions, interactions, and relations. Such deliberation by designers can be more or less intentional or conscious and often rides on a sea of tacit knowledge or assumptions of who the users of an artifact or infrastructure are—and who the users are not—and what their needs might be. Thus, the property of things and the history of their shaping over time is what constitutes, informs, and affords the "dynamic and shifting entanglement of relations."

As American pragmatist John Dewey articulates, we must be wary of placing mind and matter in opposition to one another. Matter, ultimately, is what affords and enables, constrains and impedes. In placing design in dialogue with matter, designed materiality wants to emphasize how the forms of matter are not neutral. Designed materiality centers the forms and functions that coalesce in infrastructure and the understanding that these forms are entangled in diverse histories, including histories of power and oppression, that have brought these shapes into being. This means that captured in designed materiality are also the dynamic sets of relations with the materiality of all else that comes to interact with the infrastructure: the designed materiality of wheelchairs, for example, which holds within it similarly kaleidoscopic histories of design (form, function, user assumptions), resistance, and experience. As Dewey reminds us, "matter means conditions" (Dewey 1971); as Star extrapolates, conditions mediate action (Star 1992).

In this sense, the concept of designed materiality began to emerge in my conversations with Carl when he spoke of the Victorianness of the public transport system. The idea of designed materiality then became clearer to me in conversations I had with a former government official.[3] As he spoke to me about designing for accessibility in the United Kingdom and how the accessibility regulations changed over the years, he remembered how design choices were made to change by the introduction of new regulations:

> Regulation 20a as it was for [the Rail Vehicle Accessibility Regulations of] 1998 that actually says, actually, you need to be able to maneuver in and out of it as well, because people, on paper, were designing layouts almost as if you could drop a wheelchair user in by crane, and they didn't recognize they had to maneuver in and out.

Designed materiality contends specifically with questions of how design choices that have accumulated, solidified, and been shaped within an infrastructure create differential experiences for people with different needs and expectations of the system. In this sense, designed materiality enables us to combine insights from science and technology studies, feminist new materialisms, and disability studies to speak of how congealed forms have led to congealed decisions as well as congealed behaviors, as the shapes of infrastructures also inscribe user configurations. It required, according to the government official's example above, a shift in accessibility regulations for train designers to grasp that wheelchair users cannot just be dropped into place inside a train but must board and maneuver themselves into position. The designers were making a particular set of assumptions about a new group of users for whom they had not accounted before. These assumptions had to be corrected.

Thus, when wheelchair users demand changes be made in public transportation, as Anton does when he says that he wants change rather than apologies, they are making demands for change in an infrastructure's designed materiality. In other words, Anton's demands are not solely for a change in the literal shape and form of the infrastructure. Rather, they are demands that encapsulate the experiential and epistemic. These demands for change ought to translate to shifts in "the property of things," and ought also to reverberate through the infrastructure's history, thereby also changing the various relations and entanglements contained in the infrastructure.

Designed materiality enables us to not take for granted the sociocultural histories of materials and their formations or arrangements. Ultimately, the purpose or function of an infrastructure occurs not in a vacuum but in social environments. This might be an obvious point, but if discussions of infrastructural materiality sidestep it, we end up having poorer discussions of how designed matter comes to affect some users more than others and why materiality is more salient, less ready to hand for the non-normate of the system than for the normate. Designed materiality also makes possible a toothier discussion of retrofitting as an infrastructural response to new social (and moral) pressures. Only when we consider the sociohistorical focus that designed materiality enables can we understand retrofitting to be, in Howe and coauthors' words, not only paradoxical but, ultimately, a limited response to rising demands for changes to an infrastructure's designed materiality (Howe et al. 2016).

THE PROMISES AND LIMITS OF RETROFIT

Howe and coauthors use the idea of retrofitting to describe how infrastructures "attempt to bridge timelines" and "meet new contingencies" (Howe et al. 2016,

553). The authors recognize the messiness of retrofit, discussing it as moments when infrastructures feel "sticky" or "impenetrable." Ultimately, they see and define retrofit as a paradox of both the material solidity of infrastructures (they must be at least a little malleable if they allow for changes) as well as their messy temporalities (hence the "bridging" of timelines and meeting of new demands). I explored in chapter 1 how conceptions of accessibility have shifted in the context of London transportation over the past thirty years. I showed how this change in perception brought about new provisions for access that have been limited in their success and reach. We could understand the implementation of new access solutions to public transport as retrofits as Howe and coauthors understand them. We saw the rise of new demands—disabled people's right to transportation—and their gradual integration into the system. These retrofits, the implementation of manual and mechanical boarding ramps, new low-floor buses, the inclusion of wheelchair priority areas, and more are important attempts to bridge historical circumstances (disabled people are not passengers of public transport) and contemporaneous ones (disabled passengers have a right to public transport). From that perspective, retrofits hold a powerful promise—an enchantment that comes as no surprise to anthropologists of infrastructure (see Harvey and Knox 2012; Hetherington 2014; Anand, Gupta, and Appel 2018): they promise to fix infrastructures and bring them into the contingencies of the present.

However, defining retrofits as a way to bridge timelines is to uncritically accept that they are doing that job well or that they deliver on that powerful promise every time they are deployed. To say that retrofits enable the bridging of timelines, fixing the limited foresight of the past, is to implicitly say that retrofits do this work *successfully*—that retrofits *work*, that they do, in fact, enable the patching of timelines and answering of new demands from new users. It also presumes that the infrastructures to which they are added on graciously accept these modifications, embracing and incorporating them seamlessly into themselves. That is a partial analysis: many retrofits that have attempted to include wheelchair users in London's transportation system do not operate smoothly and can themselves be sources of new problems and tensions.

Let us use, as an example, the wheelchair alert system. On London's buses, the very ones that have progressively replaced the inaccessible Routemasters, there is a wheelchair ramp siren. This siren goes off when the bus driver presses the button that deploys the manual boarding ramp. It serves to notify pedestrians and passengers that the ramp is being deployed, that pedestrians should be careful if they are walking past, and that passengers should make space in the wheelchair priority area.[4] In conversations, interviewees noted the presence of this siren as strikingly obvious when compared to the material arrangements of when a non-wheelchair-user boards a bus:

> [*Imitates alarm noise*] And everyone looks, everyone stares, and I'm like, yeah, I'm just getting on the bus. —*Alex Lyons*

Alex is a white man in his late twenties who uses a manual wheelchair on a daily basis. He has, in his words, been "a wheelchair user from a young age." He can walk short distances, and, in situations where he might be stuck, he can get out of his wheelchair without assistance and pull it behind him or use it for support. But it is more comfortable for him to not have to do so. When he wants to board a bus, however, the mundanity of the act that might be available to a nondisabled person, or any user who simply steps into the bus at the front door, is not available to him nor to any users who require the mechanical ramp. This new addition to the infrastructure—low-floor buses with mechanical ramps and their accompanying sirens—is meant to secure access for wheelchair users and smooth their entry into a bus. However, despite the provision of access via the ramps, the sirens add a layer of experience that sets wheelchair users apart even when they are constituted as possible users of the infrastructure. Indeed, this distinction is so keenly felt by another wheelchair user, Sophie, that she describes it thus:

> I don't like the fact that there's the siren that starts wailing at you, or at everybody, when you're about to get on or about to get off the bus. It's all a bit of a big faff, but you get used to it. I mean, public humiliation seems to be . . . you've got to be able to deal with it if you're disabled anyway, because people will look at you.

The siren, perhaps more than the ramp or the existence of a wheelchair priority area (which are not even mentioned in these passages), emphasizes the difference between wheelchair passengers and users who do not require it. The inclusion of the mechanical ramp on London buses may have retrofitted this mode of transport to social demands made explicit from the 1990s onwards. True, the buses from before 2005 that had steps up to the passenger space (such as the Routemasters that disabled activists locked themselves to in the mid-1990s) afforded no access. And, yes, the ramp up to a low-floor chassis has afforded physical access to the passenger area. Materially, the ramp physically enables a wheelchair user to board the bus—at a cost. The siren's activation during the ramp's deployment was likely well-intentioned: when the driver has been alerted of a need for the ramp when the blue boarding button is first pressed, the siren means the driver is not required to leave their post to keep people back while the ramp is being deployed. The siren does the job of alerting passengers on the bus as well as passers-by of a moving piece of equipment that might cause bodily harm. Yet one can easily sympathize with users who know that a loud siren accompanies them through the infrastructure; it remains a marker of difference. Never mind our associations of sirens with danger and emergencies.

That retrofitting can be a source of tension is not news to scholars in disability studies, where authors like Jay Dolmage have problematized retrofits rather than seeing them as temporal bridges.[5] A retrofit, Dolmage writes, "does not necessarily *make* the product function better, does not necessarily fix a faulty product, but it acts as a sort of correction. . . . Often, the retrofit allows a product to measure up to new regulations" (Dolmage 2017, 105). He adds, "Retrofitting is also often forced or mandated." A retrofit is not necessarily creative; it is "a sort of cure, but half-hearted" (Dolmage 2017, 105). Dolmage's analysis is in relation to disability accommodations and how they are deployed in diverse situations. While retrofitting is useful and important in the process of ensuring some disability accommodations, he argues that it is part of the logic of late capitalism and general industry trends of "temporarily correcting or normalizing disability" (Dolmage 2017, 108). Dolmage asserts that retrofitting is an exemplar of Lauren Berlant's concept of slow death, in which populations are marked out for "wearing out" (Berlant 2011).[6] Dolmage's analysis of retrofitting seems to ascribe significant intentionality to designers and architects and to characterize the choices they make in the process of retrofitting as somewhat malign. While there are surely instances of harmful intentionality in the exclusion of disabled persons from accessing infrastructures, I want to argue that what is truly complex about the process of retrofitting is that the infrastructure's own attempts at self-correcting its designed materiality are often what creates such limitations.

In placing both Dolmage's and Howe and coauthors' perspectives in dialogue with designed materiality, we can see how their arguments not only are viable but are intertwined in infrastructures. Their perspectives capture a material analysis of infrastructures that argues for how, on the one hand, retrofitting can act as an important temporal bridge, but, on the other, it is often insufficient. I argue that retrofitting, when placed in conversation with the concept of designed materiality, is a process through which the infrastructure attempts to compensate for oversights of the past without full consideration of the extensions of its own materiality into the realm of users' interactions and interrelations. That is to say, retrofits do not take into consideration the dynamic, shifting relations that an infrastructure has created and enabled over the years of its creation and consolidation. Thus, in attempting interventions to "fix" past oversights, retrofits fail to account for how the past has already shaped attitudes, behaviors, and expectations of the present. In this sense, as Dolmage argues, retrofits act as a "sort of correction," but the correction is limited if it does not also consider how the retrofit might create new sets of relations, contestations, and tensions. I will illustrate.

The new London low-floor buses, gradually deployed from 1994 onward, were designed with some understanding of access in mind. The ramps allow buses to measure up to new regulations, including the Disability Discrimination Act of 1995 and the Equality Act of 2010 (and the related Public Service Vehicles Accessibility Regulations). As I've argued, retrofitting often fails to consider the sociohistorical context in which retrofits are deployed: the ramp might enable, but the siren constrains. While the ramp enables access for a wheelchair user, the siren signals to others aboard that this is not a user like other users; this user requires a new set of relations, behaviors out of the "ordinary" set of expectations of a normate user boarding the bus. In this sense, the siren is further signaling the ways in which the bus, and the infrastructure as a whole, is still not fully aligned with the needs of these newly included users.

Understanding retrofitting as an action that is undertaken within designed materiality highlights what Leigh Star has already taught us: infrastructures do not grow de novo. With infrastructures developing along lines of generative entrenchments, their path dependency shows us that even our best intentions to retrofit may be prone to failure. This is because, in speaking about materiality, particularly when we speak of infrastructures as *invisible* (see chapter 4), we naturalize the function of an artifact or system along with its form and design. Throughout the process of retrofitting, we need to remember that we are not operating in a space of neutral materiality or of material that has not been touched. Rather, retrofits are deployed in spaces of sociohistorically designed matter that already carries values, categories, and standards. Designed materiality allows us to recognize that a designer's new goals are not unproblematically achieved. Unknown contingencies can, and likely will, emerge. Infrastructures, through their histories, have already created standard users and standard operators and has efficiently trained them in specific manners of behavior.

Such an unforeseen consequence is, for example, the continual debate surrounding access to the wheelchair priority area. Various interviewees discussed the issues they have faced in transportation due to the single space available on buses for their access. Wheelchair users interviewed often asserted their willingness to "compromise" with other users on the bus for the space and expressed their sympathy for parents with buggies:

> I'm not going to say to a mother, "Take that baby out of there, fold that." What's she to do with a baby when it's sleeping? Leave it. —*Adam*

And yet it remains a charged question. Can the two share the space? It isn't, as Alice put it to me, "very nice to shove a buggy user out of the way," but compromise between all parties is often achievable. "All parties" here includes not only the wheelchair user and other occupants of the space; it also includes the bus driver:

> But occasionally it's the bus driver who doesn't even give me the chance to negotiate with the parent in the space, they just say, "No, there's somebody in the space. You can't get on, you'll have to take the next one." My favorite phrase. —*Sophie*

Again, we see a clearly well-intentioned retrofit: low-floor buses not only have mechanically deployed ramps (and accompanying siren); they also have dedicated spaces, clearly painted blue with a wheelchair symbol and signs above reading "wheelchair priority area," and, since 2012, a public awareness campaign requesting that buggy users make space for wheelchair users.[7]

Aspects of the designed materiality of transportation have been redesigned for the benefit of a relatively well-defined user group. The material retrofitting of that infrastructure, however, does not result in an entire reconstitution of the infrastructure! The infrastructure has not grown de novo in that instance, and, indeed, it is still having to account for the embedded preferences, learned behavior, and other material constraints that it had, until the retrofit, valued and encouraged.

> The good thing is that all London buses are wheelchair accessible; the problem are the drivers. [*Laughs*] You can have the best bus, but if the driver is not willing or whatever to push a button to open the ramp, then the best high-tech bus is not worth the money, and that's exactly what happens.
> —*Kerstin*

> Physical barriers, so, we can redesign the buses, doesn't matter how much we redesign the buses, we could take all the seats out on the ground floor but if people won't move out of the way, what's the point? —*Sophie*

So, while wheelchair users are materially enabled to board the bus, other constraints of the infrastructure continue to counter that access. Interviewees often pointed to drivers not stopping when they signal at a bus stop or press the button to disembark when on the bus. This is perhaps due to drivers still being required to keep to their time schedules and fearing that the additional minute or so that the ramp deployment would take would throw off their timing. A specialist on transportation and access who works as a consultant explained to me, "You know, the traffic commissioners impose fines for late running of buses, so the drivers in turn will be penalized if they're ten minutes behind schedule. It's a difficult one to call." Similarly, Aimee expressed her empathy toward these drivers, whose jobs encompass different strains and requirements:

> I think it's probably just the pressure and the stress of the job, actually, when you're on a timetable and, you know, it can be quite a stressful job, really.

I am unsatisfied to speak of this situation as a process of finding the responsible actor for the lack of accessibility. In a similar way, Joseph Pitt, in analyzing the launch of the Hubble telescope with defects that had been identified prior to its launch, dissolves the blame game of whether it was a management or engineering problem to point instead to how the design process was what ultimately broke down. Pitt is concerned "with how the design process actually works as an infrastructure and how it can fail" (2019, 66). I would add to his exploration by asking not how the design process works *as* an infrastructure, but how the design process works *within* an infrastructure. Or, given my definition of an infrastructure's designed materiality, how the design process is *limited* within infrastructures as its plasticity reduces over time and degrees of freedom are lost in the design.

That is not to say that retrofitting is useless. Rather, it's to say that retrofitting is *limited*, for it acts within the constraints of the infrastructure and the realities that it has built over time. Though I have most frequently used the example of buses, it is interesting that these are, arguably, the more malleable piece of this infrastructure. After all, buses are replaced more often than track-based vehicles and can be redesigned much more easily than an underground station built in the early twentieth century. The process of retrofitting the Underground is therefore even more limited by its designed materiality due to its underground stations, steep steps, curved platforms, and straight trains (which are partly to blame for the infamous gap passengers are asked to mind), among other complications. This further exemplifies the power of the concept of designed materiality and how matter matters not just because of the nature of its composition and process of deterioration, nor because materiality is multiply constituted, but because all of this is true simultaneously. Designed materiality matters because it contends with how matter has been and is continuously shaped, and how it also affords particular sets of relations, actions, and interactions. It thus demands that we deal with the process of user configuration, or scripting, in which we remember how technologies' affordances create and demand specific behaviors and interactions from their users. Infrastructures are no different.

NICHE CONSTRUCTION AND AGENTIAL KNOWLEDGE

In the face of sociohistorical contingencies and designed materiality, what, then, of the work and tireless advocacy of disabled activists, wheelchair passengers, and various organizations that agitate and engage with public transportation infrastructure to be able to use it? If, as I have argued above, retrofitting is limited, my argument thus far seems defeatist. What is the use of organizing for change if change is limited? My aim here is to critically examine how, despite the various limitations of retrofitting as it is undertaken by the infrastructure, wheelchair passengers interact with designed materiality in ways that actively reshape that very designed materiality. I began to understand wheelchair users' interactions with designed materiality as constituting a process of *persistent belligerence*, following from interviews with Adam, a disabled nonwhite man in his fifties who was using a battery-powered wheelchair when we met. As we spoke about many bad experiences with transportation, he told me of a time when he knew there would be an issue with returning home on public transit, so he took a taxi that he charged to TfL. I asked him whether there was a response from TfL about the issues he faced.

> Only because I'm belligerent, [bloody minded]. I have a history with challenging and pushing them, and they know I'm a campaigner, they know I will take action, so I'm on their list as a difficult customer, handle with kid gloves, don't piss off. —*Adam*

Following Adam's belligerence, and the continual efforts of wheelchair users in their interactions with public transport, I use the concept of niche construction as a fruitful metaphor for capturing the feedforward loop between the designed materiality of infrastructures and the concrete and local acts of users. I borrow the concept from biology. In addition to nurturing the ecological approach to thinking about infrastructures (inherited from Leigh Star's work), it enriches our understandings of infrastructure. In particular, it allows us to trump top-down narratives on one end of the spectrum as well as subversive bottom-up narratives on the other to argue that the process of change in infrastructures is undertaken at both ends. Niche construction as a metaphor enables us to understand the nuances of infrastructures—their rigidity and plasticity, their evolution and stagnation, their structuring and constructed nature—and embraces the paradoxes therein without depending on them for a conclusion. The concept of niche construction in infrastructures through systems of belligerent techne allows us to dig into what creates some of the paradoxes and accept them as part of a complex picture. It further offers us optimistic paths forward for creating more democratic and socially just infrastructures.

Niche construction in biology refers to an intuitive concept, a process through which the "activities of organisms can result in significant, consistent, and directed changes in their local environments . . . substantially modify[ing] their worlds" in a "nonrandom or predictable manner" (Odling-Smee, Laland, and Feldman 2003, 5). Significantly, this process affects not only the worlds of the organisms engaged in what is also termed *ecosystem engineering* (Lawton and Jones 1995) but also other species in that environment. Classic examples of biological niche construction include beavers' construction of dams and leafcutter ants' nests. This is, however, but one step in niche construction theory. A second important part of this theory applies to evolutionary biology. Authors that incorporate niche construction into evolutionary theory argue that this mechanism is key to the evolution of species. They assert that the material changes in an environment that have been created by the niche-constructing activities of a singular species modify the natural selection pressures in the environment. Ultimately, this may influence the course of evolution for various other species in that habitat. For this to occur, however, the work undertaken by niche-constructing organisms cannot be temporary; it ought to be persistent and not erased by other populations. This persistence criterion can be satisfied by two means: either each generation of a species repeats the same behavior over time in such way that its niche is always modified in the same way, or one generation of the population does such impactful niche-constructing that subsequent generations (and other species' populations) inherit a modified environment.

The niche construction model offers an additional layer to evolutionary theory in biology, one that is not satisfied solely with genetic inheritance (i.e., natural selection) but complexifies it with questions of ecological inheritance that go beyond classic Darwinian species-centered narratives. This is, perhaps, my favorite aspect of niche construction. As a theory, it not only accepts but posits that interspecies interdependencies arise and that organisms don't just inherit generationally. Organisms inherit our worlds, and organisms modify our worlds. It makes possible a deeper ecological analysis.

The application of niche construction to the humanities and social sciences is not novel. Various authors have worked to bring the concept of niche construction to our understanding of human evolution, including through the concept of the cognitive niche (DeVore and Tooby 1987), or the human capacity of abstracting and problem solving in given situations. From there, authors have used examples spanning language development (Clark 2006) and cultural development writ large (Tomlinson 2018) to propose new understandings and readings of human intellectual history. That niche construction hasn't made a more widespread appearance in STS literature is surprising to me, as many of the concepts present in the theory, such as coupling and coevolution, are already easily comparable to ideas in STS such as coproduction. Indeed, some of the key implications of niche construction theory are that "histories of environment and organisms are functions of both environment and organism" and that "organisms and their environments are, in effect, *coevolving*, because they are *codetermining* and *codirecting* changes in each other" (Odling-Smee, Laland, and Feldman 2003, 18, emphasis mine). My best guess as to this elision is that STS scholars are generally wary of using evolutionary metaphors for fear that ungenerous readers might assume they are being deterministic or Whiggish in nature.[8]

My focus is narrower than most of the humanistic applications of niche construction. To be clear, I am not attempting to argue that infrastructures are on evolutionary tracks, nor that evolutionary theories are easily transposed to explain patterns of shifting ecologies of infrastructures. Nor am I proposing a unifying evolutionary approach to the study of infrastructures. Rather, I see in niche construction a useful metaphor to understand the specific local feedforward loops that can exist between the acts of users and the designed materiality of infrastructures. For the niche construction done by a species is always localized and contextual. In other words, it is always and necessarily situated in a particular niche, or context.

The methodology of centering marginalized experiences of infrastructure is what affords the metaphor of niche construction such interesting illustrative power. Thus far, I have foregrounded how disability activists in the 1980s and '90s leveraged conceptions of access and accessibility. In so doing, they achieved new provisions of access to the infrastructure, fighting over a century of relative stasis in the designed materiality of public transport insofar as how access was conceptualized. I have also, through historical and empirical research, shown how it is because of their actions that a slow process of reshaping the network has emerged. However, both points raise important theoretical questions. Where theories on how users matter have significantly informed the understanding that technologies are not just adopted as their designers intended and that users, too, hold power in assigning function and value to artifacts, there has been less attention to how similar (but not identical) processes also influence the shape of infrastructures. Infrastructures, as I argued above, are often read as either top-down or subversively bottom-up—in the more nuanced cases, through a continuous bottom-up "politics of life" where marginalized populations continuously work toward being recognized as citizens with rights to the public good (Anand 2017). I believe niche construction builds on these points in ways that expand current governance understanding of knowledge and expertise and that, if adopted, can enable more democratic design practices.

Illustrating Niche Construction

What, in the case of wheelchair passengers and public transportation, does niche construction look like? How does it take place, and what forces are acting? I have argued, first, that the history of public transport in London, beginning in the mid-nineteenth century, is one in which the erasure of disabled people is expected (in Carl's words, "people with disabilities didn't matter back then"). However, as civil rights movements grew, among them the disability rights movement, important victories accrued that enabled some forms of access, no matter how small and insufficient, for disabled passengers to public transport through the development of retrofitted technologies. Second, I have argued that retrofits deployed in the public transport system are limited because they contend with the infrastructure's designed materiality, not only refashioning the forms and shapes of the infrastructure but also affording and constraining particular behaviors within it.

In chapter 1 I implied that we would address questions of niche construction when I spoke of the force with which disabled people's organizations in London demanded public transportation accessibility, and how the very understanding of accessibility must have changed within the system. I invoked the work of Disabled People's Direct Action Network (DAN) and the Campaign for Accessible Transport (CAT) and their public actions, including chaining themselves to buses and stopping traffic in Trafalgar Square. These subversive actions ultimately resulted in certain changes to transportation in London, some of which had begun with the development of the Disabled Persons Transport Advisory Committee. This collective activism reaped results, but much of the benefit of these victories was diluted due to its embeddedness, still, in the social and ethical regimes of the designed materiality of the infrastructure, which remains, in many ways, Victorian.

The continual efforts of disabled passengers and organizations allow us to see new patterns of work emerging that have enabled feedforward alterations to the infrastructural niche. Far from being satisfied with the limits of the system, wheelchair passengers constantly go against it and develop their own tactics and strategies—in Adam's terms, they are *belligerent* toward the system's affordances and constraints. These truant interactions with the system, moments when wheelchair passengers bring tools (ramps, screwdrivers, bus driver's manuals) to enable their journeys, are not individual actions. They are forms of what I call *belligerent techne* that are shared and transmitted socially among groups of users who share a situated frustration. Through their actions and knowledge, they open new paths in the designed materiality. I address the full definition of belligerent techne in the next chapter. Let it here suffice to state that the belligerent techne of wheelchair passengers consists of the various ways wheelchair users interact with the infrastructure that go against its prescribed or expected use.

It is important to note that, were belligerent techne to consist solely of transient and individual actions, the work of these wheelchair passengers might not have the same type of niche-constructing impact. Recall that persistence is a key criterion of the success or failure of niche construction with regard to whether the work undertaken by a particular group of users has any staying power or influence over the environment and other users' experiences of the infrastructure.[9] Two forms of persistence are identified: one, duration over time by the same group and their tactics; or two, such significant impact caused by one generation that the changes are inherited in the niche by subsequent generations. There are, in our case, clear illustrations of both mechanisms.

Let us take the latter case: one generation's persistence resulting in another generation's inheritance. The impact of targeted campaigning for public transport access by DAN in the 1990s was such that, by the first decade of the twenty-first century, low-floor buses were omnipresent in London. A minimal level of access to these buses became expected even if, as we've seen through the limits of retrofitting, that access is still fraught. Many interviewees brought up this history of political activism in conversations, referring, even if not by name, to disabled people who "took militant direct action" (Marie) and "tied themselves to buses and locked the way of buses" (Anton). Alice, a transportation consultant and disabled woman in her fifties, had direct experience working in the transportation industry in the 1990s. She spoke of the work of DAN explicitly:

> So what happens is, disabled people raise the issue, I mean, there was a big thing called the Campaign for Accessible Transport. . . . They were the guys who chained themselves to buses and they put themselves in front of them and wouldn't let anyone . . . blocked roads and all sorts. And that was in the late 1990s and I think they were a catalyst for the organizational changes, but then it takes someone inside an organization to say, "This is not OK."

Alice's reflections on her interactions with DAN in her role within the transportation industry were also illuminating:

> One of the first things that I did was talk to the Disabled Direct Action Network because they had public transport as top of their hit list so I worked quite hard with them to make sure that they took public transport off their hit list and so that we didn't have the embarrassment of having people chain themselves to trains, which was what was happening, and I realized that not long after I'd done that that it was a really foolish thing to do, because actually what you need when you're that passionate person inside an organization is that pincer movement.

She speaks of this pincer movement as the combined pressures of campaigners outside of an industry and those within the industry who sympathize with the cause finding the momentum to effect change. Ultimately, in the 1990s, the two forces combined in such a way that a large jolt shifted some aspects of the infrastructure: new affordances, new publics, were incorporated into the system, even if in limited ways. The environment aboard buses fundamentally changed as a new space was created, dedicated to wheelchair users. The actions of that generation of disabled activists constructed new articulations, relations, and even material realities in the niche of public transport. Thus, new generations of wheelchair passengers in public transport grew to expect a modicum of accommodations and accessibility, something Alice herself articulated:

> People's expectations rise so what was OK in 1990 is not OK in 2000, and what was OK in 2000 is not OK in 2010, and what is OK in 2010 will no way be OK in 2020. Because people will have higher, and higher, and higher expectations so we as disabled people are always demanding that bit more.

The first step in the 1990s, a significant generational push that installed a new baseline of access, can, in niche construction terms, be seen as the second form of persistent behavior: one that results in such change that ancestors hand down significantly modified environments. I make a pointed choice to use the term *ancestors* here not solely because it is used by niche construction theorists, but also because it honors traditions of disabled activists who have always recognized the importance of disabled elders. As Stacey Milbern reminds us, disabled ancestors "are disabled people who lived looking out of institution windows wanting so much more for themselves" (2020).[10] This dedication to honoring the work of ancestors blurs the lines of inheritance. Troubling our conception of ancestry enables us to privilege and honor knowledge and socialities that are formed beyond the nuclear genetic family and how we use them to shape our worlds.

Thus, these crip elders, some now ancestors, reshaped the designed materiality of public transport in London. The first form of persistent behavior that can result in modifying the niche takes the shape of intergenerationally persistent demand in the form of individual and collective action. We will treat it particularly in chapter 3. The next chapter is dedicated to exploring how these systems of persistent belligerent techne fundamentally requires us to think of infrastructure not only as designed materiality but also as a space where diverse knowledges clash and infrastructures are refashioned in the process. As I argue there, crip technoscience is itself a form of niche construction: it is a form of knowing-infrastructuring that changes our worlds, unsatisfied with the environment as presented. I show how wheelchair passengers are changing their own infrastructural environments in ad hoc ways and how disabled organizations like Transport for All lobby and continue work like that of DAN. Disabled people continue to, in Alice's words, demand that bit more.

NICHE CONSTRUCTION WITHIN DESIGNED MATERIALITY

If infrastructures can be seen to shift and change over time through the niche-constructing actions of groups, why have the limits of retrofits thus continued to make themselves remarked throughout the system? Ought not the work of wheelchair passengers have successfully eradicated the barriers of access? Would the retrofits, in that case, not suffice?

I argued above that the limits of retrofit are linked to an infrastructure's designed materiality. My argument, continuing from much of the scholarship in infrastructure studies and distributed cognition, is that infrastructures, as they stabilize, go from "history into nature" (Pea 1993). Thus, when I argue for an infrastructure's designed materiality, we need to understand there to be a process of stabilization of "patterns of previous reasoning" (Pea 1993). What is stabilized in the designed materiality of London's transportation infrastructure, then, is less the function of mobility than it is the function of mobility for a particular historical category of people deemed mobile. Or, in other words, it has stabilized various sets of relations, expectations, behaviors, affordances, and constraints for those who ought to be deemed mobile, which enable that mobility.

The niche-constructing work of disabled people in transportation infrastructure is limited within the designed materiality for the same reasons that curb the impact and reach of retrofits. The actions of disabled people and wheelchair passengers, like retrofits, are working within an inherited, nonneutral, sociohistorical materiality that still carries values, incentivizes behaviors, and affords relations in ways that are not easily remolded. While the belligerent techne of wheelchair users in their daily navigations can enable their personal journeys and even, slowly, shift some of the designed materiality, this work is not completed merely by the addition of new access icons, such as a ramp to board a bus. Let me return to the case of buses to illustrate how both retrofits and the knowing-infrastructuring work of wheelchair users have to contend with the infrastructure's designed materiality.

The wheelchair priority area and mechanical ramps aboard low-floor buses in London are retrofits that respond to the belligerent techne of various disability activist groups. A specific material intervention to the infrastructure's designed materiality, it enabled an initial shift in configurations and relations in the system. The size of wheelchair areas is regulated in the United Kingdom by the Public Service Vehicles Accessibility Regulations of 2000, a statutory instrument.[11] Requirements stipulate the minimum number of wheelchair spaces on a public service vehicle ("not less than one"); the size of the wheelchair area (it must fit both a wheelchair and a wheelchair user seated in the wheelchair—though, as the former government official reminds us, even this has changed over the years); the presence, positioning, and size of a backrest; the presence, height, length, and material qualities (their solidity, surface, and color) of horizontal

FIGURE 1. The layout of the front of the ground floor of one of London's double-decker buses. The view here is from the back of the bus facing the front. An area to the right is painted a darker color with a white wheelchair symbol drawn on it. Photograph courtesy of Catherine Holloway.

handrails; the clear spaces that ought to surround the wheelchair area; and the presence of signs specifically stating, "Please give up this seat for a wheelchair user" or "words of equivalent meaning." This thorough description has resulted in several configurations of the wheelchair priority area on the various models of buses that are used throughout London's transportation system. One such example is figure 1.

There are ramps that enable access into and out of the low-floor buses. There are also wheelchair priority areas into which wheelchair users should be able to navigate. However, we have already seen that the experiential reality of wheelchair users trying to navigate into that space is much more fraught. Wheelchair users experience barriers to boarding supposedly accessible buses despite these access provisions. Let us consider some of the barriers in the context of the constraints of designed materiality. Diana, a white woman in her twenties who was studying at a London college, spoke to me about issues that she has faced with getting into the wheelchair priority area. She told me about how often that space is occupied by passengers with children in baby carriages. As she considered the conversations she has seen online on the topic, she reflected:

> I guess one of my issues is that a lot of the discourse around that particular conflict . . . gets pretty hateful. Like when I did some of the tweeting for TfL for the [wheelchair] space thing, a couple of . . . I'm thick enough skinned to not be deterred by this, but a couple of people commenting basically [said], "What the fuck does a disabled person think that they're doing on a bus, surely they understand that a baby's life is more important?"

The retrofits of the ramps and wheelchair spaces on buses has enabled easier travel for a range of users, not only wheelchair passengers. Among them are parents or caretakers with strollers, who likely also struggled to board buses due to the stairs and previously had to fold up umbrella baby carriages and carry children in their arms onboard. The wheelchair space has thus afforded new permutations of relations in the designed materiality of the bus, but it has not entirely reshaped the historical weight and experiential reality of ableism for disabled passengers. Indeed, the retrofitting of the bus has crystallized social tensions and behaviors toward wheelchair passengers, including, as Diana puts forward, the question of value of the lives of disabled persons. In the interactions that Diana has had online, where commenters question the importance of a disabled person's daily activities in relation to those of a child, societal perceptions of disabled persons that have long stood in British society (see chapter 1) are still present. While the retrofitting of buses has certainly assisted passengers with children in pushchairs in joining the infrastructure, these persons were not entirely locked out of the system in the ways that wheelchair passengers had been. Yet their claim to what is called the *wheelchair priority area* is still considered legitimate. As Robert McRuer highlights in *Crip Theory*, this valuing of reproductive abilities over disabled persons is hardly surprising to a crip analysis. Cultural signs of queerness and disability have deep entanglements in a society that privileges heterosexual, nondisabled bodies as (re)productive bodies. In the case of placing a wheelchair user "against" a caretaker with a small child, the users of the public transportation system—including, in some cases, the caretakers themselves—are accounting for a learned social hierarchy: children are more important; parents with children are more important. For the designed materiality to break out of that requires the incorporation of ways of affording new behaviors, new social values, and, ultimately, new forms of knowledge.

> I'd love some awareness-raising poster campaign, to have an adult male wheelchair user sort of, six [foot] two, sitting on his wife's lap to show, yeah,

> you can do this with a baby, you can't do this with an adult. So, fold your bloody buggy!

Faith laughed as she told me about her idea for a new awareness raising campaign that, interestingly, itself incorporated some of McRuer's notion of compulsory heteronormativity to enhance the value of the wheelchair user (an "adult male . . . sitting on his wife's lap"). Faith continued:

> More public information. But I like to take a fun approach to these things, I like to do it through comedy. I don't want to wag my finger at anybody saying, you must do this, you must do that, but show someone the ridiculousness of some things. I genuinely do think most people are unaware of the difficulties that wheelchair users find traveling around. If you show why it's important that space is free, I think it's important if you show why bus ramps need to work. I think most people will come around; I definitely do think that.

Faith firmly believes that it is generally other passengers' lack of awareness that results in such tension on buses. She, and many other wheelchair users interviewed, consistently stressed the need for "social awareness campaigns" or generally more public information. What repeatedly comes to light is that designed materiality requires more than physical shifts to encourage new societal demands and novel configurations of the infrastructure. The incorporation of new habits of thought and new behaviors requires more than the provision of what is often upheld as the ultimate accessibility icon—a ramp.

These types of awareness campaigns are being deployed by TfL, with a couple of renewals of that campaign along the way. As of 2017, two years after these interviews were undertaken, iBus prerecorded announcements were installed on buses to attempt to resolve some of the issues around freeing the wheelchair priority area when wheelchair users require it. First, as a wheelchair user boards, a recording (instead of the offending siren) states, "A customer needs the wheelchair priority area. Please make space." If the space is not freed, a second recording is supposed to be used, stating, "Customers are required to make space for a wheelchair user. The bus will wait while this happens." Thus, the London public transportation infrastructure and its administrators are coming to realize that the process of rendering the network accessible cannot depend solely on the inclusion of physical retrofits but requires a concerted, orchestrated shift to reeducate and afford new relations in its designed materiality.

As I show in chapter 3, there is good evidence that this is the result of incorporating wheelchair users' and disabled persons' experiential knowledge into the system. That is, the infrastructure is beginning to accept the manners in which wheelchair users, through their belligerent techne, are coming to (re)construct it. To identify forms of niche construction by marginalized groups as belligerent techne may therefore offer us a way of not only identifying and naming the labor of these social groups but also understanding it as desirable work, a desirable shifting of the infrastructure. To do so would require infrastructures to reckon with the consequences of their pasts beyond feeling satisfaction with retrofitting. They would have to be dissatisfied with simplistic, transient interventions to bridge the past to the present. Instead, they would have to favor approaches that ask how the past may be placed aside to entangle the present with a more socially just future.

Changing an infrastructure therefore requires shifting its designed materiality in a way that enables the incorporation of not only physical changes but also new practices, new knowledges. These incorporations, accrued through belligerent techne, result in new configurations in the infrastructure's designed materiality. When undertaken in haphazard, ephemeral fashion, the results are unequal and unevenly distributed and can result in new tensions. As demonstrated in the next chapter, tensions present in the infrastructure are often the result of clashing epistemologies and experiences. We must, therefore, tend to the ways that an infrastructure's designed materiality (as well as the tensions and relations that emerge within it) are the result of various permutations of epistemic practices.

THREE / Situated Knowledges / *"Local knowledge required."*

Diana and I met at a café on the campus of University College London in the late afternoon of a mild summer day. Our conversation began quickly; we sat down and immediately began exchanging notes about disability activism, as she began to tell me about articulations of "disability hierarchies." It took me a couple of minutes to realize that our casual introductions to one another had already dovetailed into a deep exchange on disability and being disabled in London. I fumbled to press record on my device, and we continued our conversations. Diana, a college student, born and raised a Londoner, is in her twenties and was using an electric wheelchair when we met. As we wrapped up our interview, I asked my usual final questions.[1] I began, "If I were to ask you to describe public transport accessibility for wheelchair users as it is today in London with three words?" Diana barely hesitated to gather her response: "Local knowledge required." She continued:

> My transportation travel in London becomes limited by how much exposure I've had to the part of London that I want to travel to. So, if you ask me to go somewhere in central London, my base level over the last fifteen years means that I can get myself almost anywhere in central London by a relatively effective route. If you asked me to go out of my comfort zone, to go to a part of London that I do not normally go to, so, for example, Peckham, my lack of knowledge and lack of available information means that I will not make the most effective journey [or] I may not make the journey at all. . . . When I went to visit my friend in Lewisham, I spent an hour and a half on buses from Waterloo to get to where they live. When I told them how I got there they went, "Why the hell did you do that for?! The train station's accessible." My journey back, for the portion, took me fifteen minutes. But that's because I don't have information whereas because they live there, they have their local information.

Hearing her articulate the need for local knowledge to successfully navigate the transportation infrastructure in London made me want to look through interviews again with an eye to questions of knowledge. Diana wasn't the only wheelchair user I interviewed who explicitly cited knowledge as a primary differentiator in their interactions with the infrastructure. Adam had a very similar answer when I asked how he manages to get around after he told me of various harrowing experiences in transit:

> I struggle, I only use stations that I know I can get easy access to . . . Yeah. That's how I do it. It's knowledge, you talk to other people, you say, right, you find out, you know? Social network. I'm going around here, [asking] does anybody know if there's an easy access to it or whatever? You build your own network up, which is why because of my name and who I am and what I've done, I'm not backwards in coming forwards. I do stand up and shout and people listen, people take notes and people share and we work together that way. But otherwise it's crazy.

Adam's articulation of knowledge may seem, on the surface, different from Diana's. Diana's seems to be based primarily on firsthand learning or local experiences (i.e., you know what you experience in the places you go), whereas Adam is speaking of a process of sharing knowledge. It is, in his words, dependent on "build[ing] your own network." But Diana, too, is sharing knowledge with others as her friend informs her of the accessible train station near their home. Diana's journey home was much quicker than the way to her friend's place, thanks to the knowledge shared between her and her friend. The network she has built for herself is also one that shares knowledge.

In this chapter I focus on these wheelchair users' process of acquiring, applying, and sharing knowledge and how it stands against the knowledge of the infrastructure they're interacting with. I contrast the situatedness of wheelchair passengers' forms of knowledge about the infrastructure and, specifically, their knowledge about access to it with the infrastructure's own partial knowledge of wheelchair passengers as users of the system as well as of their access to it.

In bringing together the histories of transportation and disability in the United Kingdom in chapter 1, it became easier to point to the consolidation process of public transit in London in the early twentieth century to explain the exclusion of disabled people during the formation of the infrastructure. Designed materiality, a concept I introduced in chapter 2, brought out questions of design decisions made over the infrastructure's history to point to how materiality does not, and cannot, refer solely to the material characteristics of an artifact in a singular moment but needs to take into consideration how these forms emerged and the relations that they congealed over time. Here, I exposit how to speak of an infrastructure in terms of designed materiality is to speak of questions of knowledge embedded and materialized in the infrastructure. What forms of access-knowledge do infrastructures have, and why speak of knowledge here at all?

In 1991 Latour offered a famous STS adage, "Technology is society made durable." In that well-cited article, he developed the argument that it is not enough to focus on social relations to understand "how domination is achieved" (Latour 1991, 103). Rather, he posited, it would be necessary to learn from the history of technology and to include nonhuman actors in our analyses to bypass the distinction between material infrastructure and symbolic superstructure. Using the example of hotel room keys and a hotel manager who is adamant that those keys belong in the lobby, Latour traces chains of association between human and nonhuman actors that strongly implicate guests in wanting to leave their hotel keys behind when they leave.[2] The argument continues to various conclusions, perhaps the most significant being Latour's desire to eliminate the "great divides" in order to "float upon" relativism. His concern, however, seems to be primarily to show the path through which an innovation becomes a technology, or a durable object. Perhaps, following Hughes, we might argue for consolidation of the object so that it is no longer "flexible"—that is, no alternative iterations of the artifact or system are competing. Following Latour's argument, we can posit that an infrastructure is society made durable, purely through an extrapolation of his argument from a singular technology to its application in infrastructure studies. Infrastructures are made up of artifacts and, therefore, forge durable social and material relations.

Yet this is an unsatisfying path. To constitute infrastructures as mere accumulations of artifacts and people, whether they're coconstituted through human and nonhuman actors or not, and to argue for their durability is to throw aside the stories told in the previous chapters. Relations within infrastructures change over time, and we know that infrastructures, though they carry their histories like baggage, are also somewhat malleable.

To think of infrastructures as both containing and being a conduit for various knowledges might offer an alternative path and, indeed, is what the crip feminist lens enables. I place in this chapter the work of feminist and crip technoscience scholars on the situatedness of knowledge with the work of embodied and distributed cognition scholars. As Roy D. Pea deftly puts it, "Tools literally carry intelligence in them. . . . In terms of cultural history, these tools and the practices of the user community that accompany them are major carriers of patterns of previous reasoning" (Pea 1993, 53). This follows from my argument on designed materiality: infrastructures, as the result of patterns of reasoning over time, can be understood as spatiotemporal, relational, and material orderings that are constituted through the knowledges of the members of the community of practice (Suchman 1996).

Consequently, to comprehend an infrastructure requires an understanding of the material, designed, spatial orderings of the infrastructure *as well as* the sociocultural pressures that follow from them *and* the learned, expected behaviors that are taught to members of the infrastructure's community of practice. I exposit this here as a continuation of Leigh Star's argument that infrastructures are a part of human organization and are both learned as and mutually constitutive of specific communities of practice (Star 1999). Implicit in these axioms are the knowledges held by the communities of practice who developed and formed an infrastructure and those for whom it was intended. However, I argue that we must also extend that understanding to include knowledges of competing communities of practice that use these infrastructures, whether they were inscribed in its formation and use or not. Infrastructures hold various communities of practice, and the diverse situated knowledges that thus come to interface with them may be in conflict.

In this chapter, I argue that an infrastructure's designed materiality ought to be understood as consisting of the various forms of knowledge clashing within it. Indeed, I argue that it is only through the clashing of knowledges that the belligerent techne introduced in chapter 2 can exist at all, modifying the environment in which these practices take place. It is thus important to remember that *knowledges* in this chapter are not to be left to the realm of the epistemic, that is, abstract knowledge itself. I understand *knowledge* as epistemic practices that take diverse forms. Significantly, knowledge is not something that is contained in the mind; rather, "it is something that people do together" (Gherardi 2008, 517). My aim is to show how various forms of knowledge are entangled in reified, material, and legitimized forms in the infrastructure as well as in the knowledge of wheelchair users that they deploy as they move through the system. In this sense, knowledges therefore have material consequences and are part and parcel of both the designed materiality of the infrastructure—knowledges have shaped the way designs have congealed, for example—and the belligerent techne of wheelchair users—their knowledges create strategies and tactics that, in turn, shape the infrastructure.

Knowledges emerge in infrastructure in numerous ways, from the knowledge of the engineers who designed the first underground railway, to bus drivers' knowledge of their routes and vehicles, to users' knowledge as they take out Oyster cards from coat pockets to pay for their tickets at the turnstile. My goal is to bring the question of knowledges to bear on infrastructure studies. It's necessary that we do so for two reasons. First, this will show how infrastructures are formed through partial and bounded epistemologies. I explicate how infrastructure studies can benefit from feminist technoscience theories so as to not flatten infrastructural truths as singular. This work is partially done in chapter 1, in highlighting the exclusion of disabled people from the original process of building London's transport system, but I will take it further here in showing the forms of dominant epistemologies the system offers. Second, in illustrating the tensions between the infrastructure's legitimized "official" knowledge and wheelchair users' specific knowledge of the system, I wish to discuss a long-standing question in science and technology studies about the division between lay and expert knowledge. My goal is not to reaffirm the lay versus expert divide. Instead, I question what knowledges are rendered legitimate to the system. In couching the question of knowledges as one based in recognition and legitimacy, I can therefore ask, Who is ultimately included in design processes of infrastructures, and why them?

To do this work, I build in this chapter on Aimi Hamraie's concept of *access-knowledge*, or "the historical project of knowing and making access" (2019, 5), and the concepts of *crip technoscience* and *disability technoscience*. Whereas Hamraie traces access-knowledge, in their book *Building Access*, as a sociohistorical phenomenon that shifts and changes over decades, I develop it here as an area of knowledge that is, itself, multiple, situated, and contextual. Situated access-knowledges, in their plural form, enable me to show how knowledges *about* access, about "what users need, how their bodies function, how they interact with space, and what kinds of people are likely to be in the world" (Hamraie 2019, 5), are informed by one's situation in public transportation—and that infrastructure itself holds and congeals different forms of access-knowledge. These different forms can be mapped onto what I will call *crip access-knowledge*, drawing from crip technoscience, when referring to the nuanced and complex forms of knowledge wheelchair users have of their own access needs and how accessibility works (or doesn't) for them. The infrastructure's access-knowledge, on the other hand, can be understood as *disability access-knowledge*, where articulations of accessibility are often flattened to retrofitted interventions in the designed materiality.

These differently situated regimes of access-knowledge—crip access-knowledge and disability access-knowledge—do not easily map together. Ultimately, the gap between the access-knowledge of wheelchair users and the infrastructure's conceptions of access is the space in which belligerent techne is honed. Indeed, it is through belligerent techne that wheelchair users forge access for themselves. I therefore begin by showing and describing the two regimes of situated access-knowledges described above. The former contains depths and layers of *knowing-how* that are deeply entangled with its firsthand experiential nature. The latter struggles to encompass the same richness; the public transportation infrastructure's access-knowledge is often limited through the social history of its designed materiality. I argue that the access-knowledge of infrastructures is thus not just partial but unfortunately also limited, as it has to translate complex access-knowledge into infrastructurally legible checklists, standards, and regulations. Having explored these two forms of access-knowledge, I focus then on the concept briefly introduced in chapter 2, belligerent techne, to show how these forms of knowing-making developed by wheelchair users are honed against the infrastructure's access-knowledge. In a fascinating cycle, these two forms of access-knowledge are constantly reshaping and reinforming one another. I finish the chapter by thinking through how we can ensure that this reshaping is being done in feedforward loops for continual niche construction of the infrastructure, rather than through frustrated loops of new belligerent techne that should, instead, be integrated as infrastructurally productive techne.

CRIP ACCESS-KNOWLEDGES

In what follows, I lay out wheelchair users' particular form of situated access-knowledge. As I will illustrate, wheelchair passengers who navigate the world of London's public transportation hold complex and nuanced crip access-knowledges. To travel in London as a wheelchair user requires layers of planning and problem-solving abilities, strategies, and tactics. I thus offer what I have termed a *blended narrative*, the result of my analysis of the various interviews turned into a collective story of what emerged as recurring themes of how wheelchair users make their journeys. This is a way I found of doing justice to the stories that were generously shared with me. Following from Oliver's methodological framing of "[illuminating] the lived experiences" of this group of users, this story-like approach highlights the access-knowledges developed by wheelchair passengers within London's transport network.

When I asked some of the wheelchair passengers interviewed for this research what an average journey was like in London, Diana and Kerstin, interviewed separately and separated by their age and nationalities (Kerstin is from continental Europe; Diana was born and raised a Londoner) gave me the same answer: "There's no such thing as an average journey." The lack of predictability in interactions with infrastructure, including its myriad passengers, the staff on buses or at stations, and everything else in between, means that wheelchair users acquire and learn many skillful ways of coping with their non-average journeys. Or, as Anton put it to me:

> I suppose one trick is to get to know the transport network intimately because you can't just pop out of the house and say, "I'm going to town." I have to think it through. So I know the bus network extremely well. I know the accessible Tube station list well. I know lifts. Knowledge. Knowledge is a trick.

Knowledge of the system was brought up multiple times in my conversations with wheelchair passengers. It is spoken of as an enabling factor that "makes a difference" (Alan) for getting around. I thus examine here the layers of wheelchair passengers' access-knowledges of London's public transport.

Our story of navigating the system is composed of four moments for wheelchair users: the decisions that set the tone for all future interactions with the infrastructure; the planning that occurs before they set out on a journey; the journey itself; and, ultimately, a dedication to acquiring more refined access-knowledge through collaborative, extended, and embodied knowledge sharing practices. These moments offer examples of what Aimi Hamraie first articulated as crip technoscience and further refined with Kelly Fritsch. The notion of crip technoscience is one that resists the bifurcation of defining some knowledges as expert and others as lay because it posits that disabled people "are experts and designers of everyday life" (Hamraie and Fritsch 2019). In this sense, my articulation here of the access-knowledges of wheelchair users is only possible using the crip feminist lens, which highlights the knowledges of disabled persons as not only legitimate but desirable forms of knowledge. Hamraie and Fristsch contrast crip technoscience with disability technoscience, a regime of knowledges *about* disability and disabled people often predicated on traditional expert relations. In other words, disability technoscience strips agency and knowledge from disabled persons and is easily satisfied with expert knowledge about disability rather than the experiential knowledge of disabled persons themselves. This section is dedicated to the former, highlighting how wheelchair users' crip access-knowledges are world-building strategies through which they become experts on and designers of their everyday lives.

Setting the Terms of Engagement

While describing where she chose to live in London, Aimee told me, "You just have to make so many different decisions if you're disabled." She had been speaking about how she had weighed a list of criteria: desirability of location, cost of housing, accessibility of housing (both physically and financially), and proximity to accessible and well-connected transport. When she first moved to London, she did not factor in the latter and lived in a location that "was nowhere near an accessible Tube station." So she moved, two years later, to a more expensive location near an accessible station with a direct link to where she works.

This is one of the first steps in navigating transportation for wheelchair users: a series of decisions that must be made long before they head out of their front door . . . including about the choice of location of the front door! Wheelchair users are already geographically constrained in London given the low number of accessible stations and their poor connections to different lines. As a result, interviewees pointed out a trade-off between living in an inexpensive neighborhood but spending more on transport, or living in a more central location, closer to work, but paying less for transport. As Alex Lyons described it:

> Most people move to London and think, "Oh, I'll just find anywhere," and they very often find rooms at the top of houses with three flights of stairs in zone 6 because it's cheap and they don't mind the travel, because they just jump on the nearest Tube. Well, when you can't jump on the nearest Tube because it's not accessible, you have to really think about it. So I literally looked at a Tube map and was like, I'm going to house hunt in this area, I'm going to house hunt in this area, I'm going to house hunt in this area, because for me there has to be an accessible Tube station nearby.

Not unlike Alex Lyons, Chiara, a multiply disabled white nonbinary person in their midtwenties, described how they decided to move out of their parents' home because it had poor connections to a job they had secured:

> [My job] was based in Tufnell Park and I'd previously been living in Chiswick . . . and it took me something like an hour and fifteen minutes, and transport in general takes a lot out of me, I find it very stressful. And before I had a wheelchair there was also the matter of standing up when I was walking unassisted, not being able to get a seat. So . . . I moved to Tottenham, which is much closer by. It would have taken me about forty-five minutes max to get to work.

Chiara mentions a "before"—before they decided to acquire a wheelchair. Their choice of acquiring a wheelchair for their disability was a significant step for them, as it is for many wheelchair users. A few interviewees discussed having connective tissue disorders that affect their joints, often resulting in joint dislocations that cause walking to be difficult and painful.[3] Due to these disorders, these interviewees all chose to buy wheelchairs for mobility purposes. Many of them described the wheelchair as a liberating mobility aid that, once acquired, allowed them to cease taking pain medications (Char Aznable) or to spend a day out with friends without worrying about becoming tired (Alanni). At the time of our interviews, two of these interviewees, Chiara and Alanni, had only been using wheelchairs for a short time (up to six months). Some days, they still chose to use crutches rather than their wheelchairs. For Chiara, these days were described as useful, used to scout out accessibility for when they use their wheelchair. Alanni, on the other hand, does not self-propel and uses a wheelchair on days they are out with their partner.

The use of a wheelchair was an important factor in this research as it is concerned with the accessibility of public transport for wheelchair users specifically. Yet choices are not only dependent on whether one will use a wheelchair for the day. There are many types of wheelchairs to choose from, with the more common ones being electric (power chairs) or manually propelled (either self-propelled or pushed by a personal assistant or caregiver). There are also types in between, such as power-assist chairs that have motors to add a boost on difficult ramps, for example. Choosing between types has consequences for transport accessibility, given that power chairs are generally heavier and therefore more difficult to move manually. Some wheelchair users opt for a single type. Peter, for example, chose a manual wheelchair, while Sal chose a power chair. For Sal, it was a question of independence:

> If I'm in my nonelectric wheelchair, my carer can tip the chair and get me up onto the floor of the train. But it's just not possible with the weight of the

electric wheelchair. And that is the difference between me being independent and not. . . . I don't want to have to rely on having a carer to get around London.

Peter, on the other hand, said that he would require a personal assistant regardless of what type of wheelchair he picked. He therefore chose a wheelchair with wider and bigger wheels that would allow him to self-propel on flat surfaces, knowing that in cases where he might become stuck, his assistant would be there to help him. The lightness of the manual wheelchair also allows him better mobility as he teaches his personal assistants how to easily maneuver him over gaps and down some steps:

> I need an assistant with me for lots of things anyway. . . . I think an electric wheelchair sort of hinders you in some way, it gives you some independence but the reality is I'm going to need help at the other end of where I get to anyway, so the person who comes with me might as well push, and that allows me to sort of cheat, because I can get up curbs, I can get up a couple of steps, even a flight of stairs.

In other cases, both types of wheelchairs are kept as possibilities. Michael J., for example, owns one of each type and finds himself having to choose between them before he leaves home. On the one hand, the power chair allows him to save his energy; he does not have to propel himself and can use his energy for other things. However, he finds that using public transport on the power chair is more difficult.

> As much as it helps me in many respects, having an electric wheelchair to get on and off buses is an absolute nightmare because I need so much room to be able to turn around to be able to get into the wheelchair space, and you know, people can be a bit rude sometimes. . . . If I know I have to use public transport, I've realized in a very short time that I could not use my electric chair, so I have to use my manual.

Having selected their wheelchair and knowing it will affect their interaction with the infrastructure leads these users to think measuredly through their transportation options. Wheelchair users discussed how they had to consider whether they would acquire a private vehicle either as their primary or as a supplementary mode of transport. For example, Um Hayaa, Basil, and Sophie all chose to invest in cars in addition to using public transportation. Um Hayaa, a North African woman in her midforties, decided to buy an adapted vehicle because she felt that she was spending too much time on public transport to take her daughter to extracurricular activities, as she had to take three buses in either direction. Nearly all interlocutors described using taxis as an important part of public transport but often spoke of them as an alternative to modes of transport such as buses and the Over/Underground.[4] All boroughs in London have a TaxiCard scheme, which provides subsidized taxi or private hire service for "London residents with serious mobility impairments or who are severely sight impaired" (London Councils n.d.). However, this subsidy is limited, and it was often pointed out in interviews that taxis are prohibitively expensive to use as a daily habit, and my interviewees mentioned experiences of drivers ignoring them or not picking them up. The taxi option was often described as a last resort, used when all other options would take too long, require too much energy, or completely fail their needs.

> Once before we've had to call a taxi because we were going to be late for our appointment because the buses will not stop. *—Leda*

> So I usually end up having to use taxis because the bus routes don't really go where I need to go. *—Linda*

Interviewees spoke of a multitude of issues in relation to taxis, including taxi drivers who would refuse to take them on or say that the ramp on their vehicle is broken or unavailable. I will not expand too much on the topic of London cabs, however, in order to not overextend the scope of this discussion. Suffice it to say that many of the barriers that have been discussed in chapter 2 often apply to wheelchair users' use of London's black cabs as well, despite Transport for London's claims that all London cabs are fitted with ramps and are fully wheelchair accessible. They are not the seamless solution that they appear to be.

Having established the terms on which one will engage with public transportation, wheelchair users can then begin in earnest to think about their journey to come.

Planning One's Journey on Public Transportation

All interviewees placed a strong emphasis on planning throughout our conversations. The resources to do this planning are multiple, the most official source being Transport for London, which provides a route planner on its website and produces an assortment of maps that are available both online and at most stations. Beyond the popular schematic Underground map, two others are important to this planning process: the "Step-Free Tube Guide" and the "Avoiding Stairs Tube Guide." The former, discussed in more detail in the next section, provides the user with detailed information about accessibility at all stations but is also criticized by some interviewees as being anywhere from "a bit" (Kerstin) to "incredibly" (Chiara) complicated.

Wheelchair users spend a significant amount of time trying to "work [out] the route" (Marie) to ensure that it is accessible. Aimee estimates that she probably spends "an hour a week" looking at the TfL route planner. The work involved in planning out routes requires several modes of research on the part of these passengers, who know that problems along the way can result in being late at best or getting stuck in an inaccessible station at worst. So they plan. Many interviewees spoke about developing a "mental travel map" of the system, as D put it. This is a version of the network that they know is accessible to them, personally, as opposed to what the official maps show. These mental maps are developed through experience, trial and error, and research and observation. They can be condensed into a collection of reference points for routes or areas that are less problematic to their users. Alanni, for example, speaks of "safe routes," an adjective also used by D. The strength of these users' knowledge of the system is seen in interviews through the way that they confidently recite their routes and stations of preference, or which ones they know are out of bounds to them. Their knowledge is such that it will sometimes clash with that of TfL employees, as they are aware of places where they can change lines despite the station not being marked step-free.

Alice, likewise, has mental maps of the network that she depends on. She reflects on some of the limits of this kind of map, however, saying that, whether they are mental maps materialized onto paper by the user or reference points in one's mind, they do not easily keep up with changes that occur in the system. There is, she said, a "lag" between the provisions of access in the network and wheelchair passengers' mental maps:

> I still have network patterns in my mind, network maps, that actually may no longer be valid. Not that they don't work anymore, but more would work and I could make more connections and do more things.

In these situations, more access-knowledge in the form of research or experience is required to ensure that these personal maps are kept up-to-date and include any malfunctions that may be occurring in the system just before one leaves home. For this, the use of smartphone applications or social media to check for updates is necessary. One of the most frequently mentioned Twitter accounts is "Up Down London" (@TubeLifts; also a website), which collects and collates information from various TfL outlets on the status of lifts at Tube stations. Additionally, TfL has a Twitter account specifically for accessibility concerns (@TfLAccess), which also provides up-to-date information on the network as well as guidance on where to look for additional information and help (e.g., phone numbers, emails, and links to access guides).[5] Among the interviewees, three mentioned checking these social media outlets as one of the things they do before heading out the front door.

> I check moments before whether the lifts are usable. —*Anton*

Beyond the logistical preparations, planning out routes and checking whether lifts are in working condition, there is an emotional preparation that takes place as well. Interviewees often referred to their emotions while traveling as negative ones. Faith describes a constant "anxiousness," Marie says she becomes "nervous at the thought" of leaving the house, and Anton speaks of the experience being "stressful," among other examples. The emotional preparation includes having to "psych" themselves up to leave the house (Marie), which works against the feeling that they can ever be spontaneous about their journeys.

> [U]nless you get in a cab, you can't do that. You have to plan ahead, well, that's going to take that long, and it's either this route or the other route. —*Alex Lyons*

It is a circular process through which logistical planning might help in gaining confidence to leave the private sphere. This, however, is at the expense of spontaneity, of traveling whenever one desires. Consequently, despite having made fundamental decisions about where to live and preferred assistive technologies, wheelchair users also must dedicate time and effort to learning the ins and outs of the transport infrastructure. Yet even then, a smooth journey to their destination is not guaranteed.

Facing the Journey

Even armed with access-knowledge about which stations are wheelchair accessible and what connections are possible, wheelchair users still face barriers in London's transportation system. Research and preparatory work does not always suffice to ensure a pleasant or hassle-free journey.

> We normally plan, if we go to a place that is very far, I plan the journey beforehand. Planning it doesn't mean that it will be smooth; there'll be problems. *—Um Hayaa*

Wheelchair users have therefore developed various techniques for dealing with the barriers encountered along the way: tactics that have been developed over time through experience or through conversation with other wheelchair users. I offer here a handful of tactics wheelchair passengers spoke to me about to illustrate the extent to which access-knowledges are not only abstract knowledge that occupies the mind but are forms of engaging with and molding the infrastructure itself.

Emotional states—a friendly demeanor, feeling confident—were often mentioned as key strategies for traveling. Alan, for example, discussed in our conversations the importance of developing a rapport with station staff, especially at your local (and most often used) station:

> People like personal relationships and . . . you benefit from personal relationships. I . . . got off the train at Richmond on Sunday, and the guy that got me off the train, I got to the bottom of the ramp, and he stopped me and he said, "Hey! Got a bone to pick with you! I saw you at Kew the other day and you didn't say hello to me, what's going on?" It was the sort of conversation that you have with a mate, you've been in town and whatever and you've not seen him. So it really is . . . you go about friendships with people that dramatically improve the experience of travel. *—Alan*

Friendliness can make opportunities for wheelchair passengers to speak to people and ask for services or favors. It can work in conjunction with confidence, which was also cited as necessary and useful in a variety of ways. Confidence can enable one, while traveling, to vocalize one's needs, to be "happy to press emergency buzzers and buttons and speak to station staff" (Alex Lyons). Traveling at all requires some level of self-confidence.

> Basically, you can't be scared when you're using London Underground, and if you're a wheelchair user, these two things don't mix up because otherwise you basically can't use the Tube. *—Kerstin*

This is particularly true when traveling alone, when one might have to resort to asking strangers for assistance in case staff members with manual boarding ramps do not show up for disembarking or boarding. Many stories were shared with me of having to ask other passengers to assist with the actual disembarking or by going to search for a staff member for help.

> I've had the same experience with . . . being put on the train in central London and not knowing that someone's going to be there to get me off the other end. At worst, [I] ended up with a couple of drunk football supporters literally picking the wheelchair up and carrying it back onto the platform.
> *—Basil*

Confidence can also enable people to stand up for themselves in cases of confrontation, though it places the onus on wheelchair users to force their way onto a bus or a train and, again, vocalize their needs.

> It's not that easy, it's not. Especially when you're a person who doesn't like confrontation. Often you'll just back down, just wait at the bus stop and wait for another one to come along. *—Leda*

The barriers to transportation access are ultimately felt through their cumulative nature, where one might have to wait through multiple buses or trains to be able to board at all. In these situations, Adam pointed out that it can pay to be "belligerent" and make use of what, thanks to his term, I've come to call belligerent techne:

> We're always coming up with ideas. When somebody says, "you can't do that," one, *can't* doesn't exist in the English dictionary, and two, watch me. Once I've done [it] . . . You were saying? "Can't get the ramp out for you." I'll crawl on. "No, no, no, you can't do that." Why not? What do you expect? I stay here and don't do what I want? No, I'll do it. I'll do what I'm comfortable with doing. Many people will do so. *—Adam*

Crawling on, as Adam and Aimee describe, is one way of getting around barriers, or a last resort if a ramp isn't working along the way. When faced with a gap or a step, other wheelchair users described their own ways of surmounting them. This is Alex Lyons's:

> So there's a gap, so I get out of my chair, step down onto the platform, stand there, hold on to the train, pull my chair behind me. *—Alex Lyons*

Passengers like Chiara, Alanni, and Leda, who have the ability to walk short distances, will stand up from their chairs and walk, using the wheelchair as a crutch. Alternatively, they might have a walking stick or crutches hanging on the back of their wheelchair, which is helpful in these situations.

> Then I would have to get the stick and stand up for a bit, and get on the train, then Simon lifts the chair up and I sit down again. —*Leda*

Other ways of getting around barriers include balancing on back wheels and "bunny-hopping" down a step or even a couple of steps. These abilities, which are also called "wheelchair skills" in some training courses, can be self-taught or acquired by attending specific courses. Very few charities in the United Kingdom provide wheelchair users with wheelchair skills training courses, but I was fortunate to have been allowed to observe a session held by one of them. Interestingly, none of the wheelchair users interviewed for this research had attended such courses; all were self-taught.

> So, you know, jumping a wheelchair across a gap is just something I've taught myself over the years. —*D*

It is important to point out that those who resort to these approaches are all wheelchair users who use manual wheelchairs, as these skills are often prohibited by the weight of power chairs.[6] Having these skills is also dependent on the wheelchair user's type of impairment and strength, a point that is not lost on any of the interviewees, who, after stating that they are able to do these tricks, often continued with a caveat:

> It's like that, making do and getting on with it. Someone like me will be able to do that, but other wheelchair users won't be able to do [it] because they can't get out of their chairs without help. —*Alex Lyons*

As a result, wheelchair skills give those able to use them some flexibility in the face of ad hoc barriers. Most of the cases where interviewees mentioned balancing on back wheels and bunny-hopping were situations where something had not gone as planned: an unexpected step, a ramp breaking down, a staff member not being there to deploy a manual boarding ramp, or the wheelchair user ending up in the wrong carriage.

However, wheelchair skills are not always a feasible way of tackling issues while traveling, particularly for electric chair users or those who might not have the ability or confidence to use such techniques. Marie, for example, called wheelchair skills "stunt riding, rather than general everyday riding." She says she could not do those things as her condition physically prevents it, affecting her "internal core structure," joints, and muscles. Some wheelchair users who do not have these specific skills can use other techniques:

> If somebody doesn't meet me at the other end, I'll stick my footplates in the door. The door doesn't close, the train doesn't go anywhere. *—Alan*

> But if they've [the bus drivers] just ignored me, or they've tried to drive and I've heard the buzzer go, or if they've seen me and they've made eye contact with me and they know I'm waiting for them, if it's safe, I'll drop down the road in front of them and just sit there. I did this recently. That's how I stopped this guy leaving. *—Char Aznable*

Both speakers above are white men. Mentions by women and nonbinary persons of using themselves or their wheelchairs to block closing doors, stop buses from driving away, or "scatter a crowd" (Anton) are almost absent from interviews. This suggests a gendered distinction in what physical strategies people feel empowered or comfortable enough to risk, including the level of disruption to the infrastructure that they are willing to enact to ensure their own access. For example, when asked whether she had ever used her chair to block doors, Faith responded:

> No. However, because I wouldn't do that, because I wouldn't want to damage my wheelchair or me.

She did have an addendum, remembering she had once blocked the doors, but only because she did not have another option besides staying on the train until Wolverhampton—nearly an hour out of her way. It is interesting to note the different levels at which a wheelchair user might feel that it is necessary to use themselves as a tool or, ironically, as a barrier.

To get around some barriers, Alan told me that he and his partner had developed an idea:

> One of the things that we've developed over the last three years is a tool kit of things that we take with us.

This tool kit includes a wide range of literal tools and other useful items: a T-key (a square spanner used to unlock the manual boarding ramps from their holders), a screwdriver (to fix and unscrew ramps along the way), a foldable

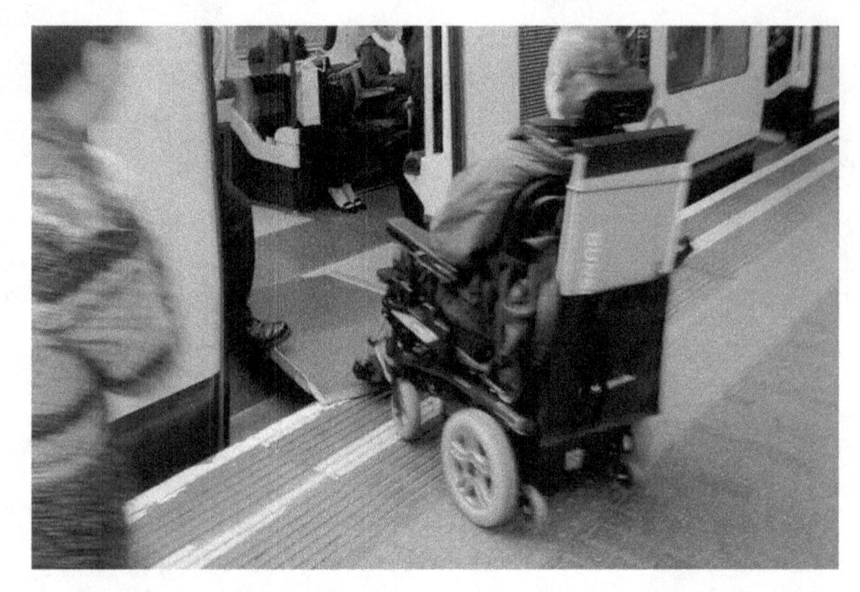

FIGURE 2. Alan boarding a London Underground train with the assistance of a manual boarding ramp deployed by station staff. In a pocket on the back of his wheelchair he carries another folded-up ramp. Photograph courtesy of Alan Benson.

lightweight portable ramp (carried on the back of Alan's wheelchair, see figure 2), and a copy of the Big Red Book, as the TfL driver's manual is known. While other interlocutors may not have used the term *tool kit* explicitly, they similarly carry useful items to fix issues during their journey, be it their own boarding ramps, the driver's manual, or other items.

> I've been out and bought a two-foot ramp so I'm going to have a ramp on my wheelchair. If I do get somewhere and I need help, I can just have somebody to hold the ends while I get up. —*Marie*

Anton mentioned having the manual on his phone, whereas Adam quoted from it during the interview ("The Red Book says you have to give up that space"). Alanni, however, did not come by the book on their own. They told me about how a friend of theirs had sent them a copy of it ("especially the pages relating to disabled passengers") because the friend thought it might be helpful. This connects us to the last, but perhaps most important, aspect of wheelchair users' crip access-knowledge: they do not come by this knowledge by themselves alone, though personal experience definitely adds to the breadth of their knowledge.

Rather, these forms of knowledge are distributed and shared among diverse and vast networks of disabled persons and allies.

The Interdependence of Crip Access-Knowledges

We observed in the first chapter the history of disability activism in the 1980s and 1990s in the United Kingdom and how it pushed for and succeeded in achieving some access victories, including low-floor buses and accessible Underground and Overground stations (however limited in number).[7] This collective work has not ended, nor has it been the sole form of the social dimensions of knowledge that wheelchair users and disabled persons have forged over the years. The sharing of access-knowledges among wheelchair passengers is done in formal and informal contexts; here, I avoid drawing distinctions between levels of formality. In so doing, my goal is to show how the access-knowledges of wheelchair users are not the result of personal experiences that accumulate into practical and skillful coping abilities. Rather, they are dependent on each individual's experiences and acquisition of knowledge and are *also* the result of a collective sharing that produces layered, nuanced, and contextual access-knowledges.

This collective sharing can be as simple as instances already described: Diana, in going to visit her friend in an area of London she is not personally familiar with, learns from her friend that a more convenient station does, in fact, have access provisions. Alanni's friend gives them a copy of the Big Red Book, thinking they may be able to make good use of it. As Adam said, when you encounter an issue, "You talk to other people, . . . you know, social network." Um Hayaa described her experience with this and the importance that her personal networks have had in supporting her, particularly in relation to learning how to navigate the world in her wheelchair:

> I had already friends who are disabled and wheelchair users even before I became a wheelchair user, because of my condition since birth, a progressive condition, and because of my sister who became a wheelchair user, I think eight years before I became a wheelchair user, so I had a network of peer support who are disabled and who are wheelchair users as well. We discuss these issues, I speak to my friend, How do you do this, how do you do that? And see whether this technique is going to work for me or not, so I think you know how your body works, you know how your mind works, you know your personality, you know your limitations. And everyone has limitations, able-bodied have limitations, no one is powerful, no one can manage

> their life from A to Z without asking for help from the other. We will never be able to reach anything if we don't share, if we don't do things together. . . . So yes, as a Muslim believer I believe we need to maintain good relations with others, we all need each other, we all have limitations in our thinking, in our perception, in our hearing, in our understanding, and we learn from each other as well.

Um Hayaa deftly expands her circle, speaking from her particular experience of sharing knowledge with her peer circle and speaking of how necessary that sharing of knowledge is for everyone. She wraps her personal identity as a Muslim woman into her belief that this sharing is not only a question of maintaining "good relations" but of interdependence as a value. She was not alone in placing interdependence as a key aspect of this sharing of experiences and knowledges. I mentioned, in chapter 1, the existence of a nonprofit organization (in UK parlance, a charity) called Transport for All (TfA) that has continued to advocate and agitate for improved access for disabled and elderly passengers on London public transport. I offer their work here as a primary example of how disabled passengers lean on a politics of interdependence that both invigorates and, more importantly, shores up the complexity of crip access-knowledges.

Transport for All as an organization was born out of Dial-A-Ride and TaxiCard user groups, two initiatives for providing transport to disabled people that have been critiqued by disability activists for their segregated nature. Transport for All has widened the previous goal to demanding improved accessibility in all modes of transport for elderly and disabled passengers. It has led or had close involvement with campaigns to demand improved infrastructure, additional funding, and maintenance of subsidized transport schemes.[8] Beyond leading and participating in campaigns, this charity provides media support and plays a role in lobbying and liaising among politicians, industry, and members of their organization. I offer here but a few examples of their work to demonstrate the value of interdependence for crip access-knowledges.

Transport for All, and various other disability advocacy organizations and disabled people, come together, especially on social media, to collectively share knowledge and experiences to enhance one another's journeys and to shame companies when they face accessibility barriers. Several interviewees mentioned using Twitter as a way of adding pressure on and demanding responses from companies.

> Shaming, public shaming, is important, and Twitter is brilliant for that. —*Anton*

Pulling on that resource, Channel 4 News developed a series in 2013 called *No Go Britain*, where episodes featured disabled people's experiences of using public transport throughout the country, in London in particular. In one episode, *No Go Britain* asked disabled people to take to Twitter to recount their experiences, and the hashtag #NoGoBritain is still used as a way of tagging inaccessible transport experiences in the country.

But social media works for more than sharing negative experiences, though that alone is useful in letting people know where not to go. It is also an effective way to diffuse knowledge of where to go, how to get help, and how to report errors and incidents. Some social media accounts have been created exclusively to share up-to-date information about the status of the transport infrastructure in ways that are important for disabled passengers. An example of this is @TubeLifts (or Up Down London), mentioned above, a bot that pulls from TfL data to tweet when an elevator is faulty at a station so those who need elevators know not to depend on that station for their journey. At least one member of the team that developed the account is a wheelchair user themselves.

Transport for All leverages many of these approaches, retweeting cases and collating stories of negative experiences, sharing travel planning tips (even providing an advice hotline), and helping log complaints to TfL and other transport service providers.[9] Collective complaining was mentioned by various wheelchair passengers as a necessary step in improving the system:

> I want them to see statistically just how bad things are, and if I can contribute to that, then it's helpful, I think. —*Anton*

This collective complaining can be seen as a form of forced mapping of issues. Anton isn't speaking of it as a way of instantly improving infrastructure, as if a complaint would have a magical impact on the problem. Rather, the influx of complaints is a long-term strategy, whether it is done through social media, emails, or Transport for London's own logging system. It provides undeniable evidence of the problems that wheelchair users are facing, including details of when, where, and how many times they encounter a problem. It's a manner of making issues visible to an infrastructure whose knowledge of disabled passengers' needs is, as we will see, rather elementary and essentialized.

Another Transport for All strategy is to offer politicians and transport industry professionals accompanied trips with disabled passengers. These trips consist of an afternoon using public transport with disabled and elderly members of the organization, making politicians, for example, witness firsthand the barriers their constituents encounter on journeys. Some of these trips are documented on Twitter with the hashtag #getyourMPonboard (get your member of Parliament on board).[10] Transport for All argues that these trips are important as "there is no substitute for the real experience of travelling with an older or disabled constituent to see the reality of transport access" (TfA, 2017). Alan, for example, accompanied Zac Goldsmith, then Richmond Park MP and Conservative candidate for London's mayoral race, on an afternoon in 2015. On that trip, Alan said things "turned out quite well," because assistance showed up with the manual boarding ramp but put him in the wrong carriage, meaning he was unable to disembark at Green Park as their carriage was not level with the accessibility humps on the platform. In the case of accompanied trips, TfA members are hoping that things *do* go wrong, to better illustrate the barriers they come up against.

> Fortunately, we'd taken our own ramp, so actually I was able to get off, but it demonstrated that even when things are supposed to go well and everything is set up to go well, it can go quite badly wrong. —*Alan*

Using his own devices to create access only further illustrated to those traveling with him the extent to which current provisions are not enough.

In this approach, Transport for All is effectively getting politicians and decision makers to engage with the crip access-knowledges of disabled passengers, which sometimes, as when Alan uses his own manual boarding ramp, go directly against the access-knowledge of the infrastructure (which placed him in the wrong carriage). These can be surprising actions to see from wheelchair users, often presumed to be passive users of the system. They demonstrate to politicians that wheelchair passengers are not idle users of transport and are willing to go to extraordinary lengths to deal with the problems they encounter. The trips, then, provide direct experience, showing the access-knowledge that disabled passengers have to those who have some level of control over allocation of budgets to improve the system's issues.

In these instances, Transport for All occupies an interesting position in the crip access-knowledges network in the ways the organization interacts with disabled Londoners. They position themselves as a central node in collecting and distributing various crip access-knowledges. These interactions do not depend on the mediation of official transportation institutions. Rather, their strength lies in being outside of those systems, creating alternative paths for connection that are not constrained by the infrastructure's own access-knowledge. Borrowing from the work of Rabeharisoa and coauthors, we can name Transport for All in this context a *representative organization*—they are, and act on behalf of, specific represented people (disabled and elderly passengers) and bring their members' concerns to light (Rabeharisoa, Moreira, and Akrich 2014). They collect and also generate experiential knowledge (e.g., personal experiences) and, as an officially registered charity in the United Kingdom, leverage institutional legitimacy. This legitimacy assists in having their concerns listened to by other legitimized forms of knowledge.

This is by no means an exhaustive narrative of the variety of decisions, plans, tactics, tricks, maps (mental or otherwise), or tool kits that collectively constitute the layered and nuanced crip access-knowledges of wheelchair passengers. Rather, the purpose here is to portray the amount of thought, effort, creativity, and ingenuity that is required from wheelchair passengers to travel in this system. The blended narrative aims to show the types of decision making and ad hoc problem-solving skills that are developed. This demonstrates the extent of knowledge that these passengers have accumulated not only on access but also on the infrastructure itself. As I show next, this stands in contrast with the infrastructure's own understandings of access.

DISABILITY ACCESS-KNOWLEDGE

In the previous section, I showed through a blended narrative the extent and nuanced nature of wheelchair users' crip access-knowledges. The process of decision making and strategies used to navigate transportation speak to an ability to deftly maneuver among diverse emotional states, create useful representations to navigate the system, and share these knowledges with others who may make use of them. The complexity of these crip access-knowledges stands, I argue, in stark contrast to the relative flatness of the disability access-knowledge of the infrastructure. The disability access-knowledge of public transportation in London, and those inscribed in it as historically imagined and normate users or maintenance staff, is marked by its simplicity, originating from both its cultural history and the language of infrastructures themselves—their standards. In this sense, the concept of disability access-knowledge builds on Aimi Hamraie and Kelly Fritsch's description of disability technoscience. Hamraie and Fritsch describe disability technoscience as a field of "traditional expert relations" that designs *for* disability and disabled people. In this form of knowledge, disabled people are perceived solely as users to whom solutions need to be offered by experts who truly understand the issues at hand.

To understand infrastructures as knowledge is also to understand infrastructures as partial knowledge—namely, the partial knowledge of those who have built and maintained them over the years. Here, I argue that an infrastructure is, at least in part, the materialization of the situated knowledge of its developers and maintainers; it is the carrier of patterns of previous reasoning. These previous reasonings are materialized in infrastructures through what we have come to understand, variously, as scripts (Akrich 1992; Akrich and Latour 1992), or configurations of the user through texts (Woolgar 1991).[11] Through these scripts, designers of technologies create prescribed uses for their artifacts, nudging users toward particular behaviors. We see various examples of this throughout the transport network—some in literal scripts (such as signs or maps) of the system, others in the designs of buses and trains. The wheelchair priority area and the step-free map are examples of the disability access-knowledge of the system. I purposely opt here to use examples where conceptions of access are already at the forefront: both artifacts allegedly enable wheelchair users to navigate the system. But, as we will see, the infrastructure's disability access-knowledge ends up creating new permutations of inaccessibility due to its limited understanding of accessibility itself. These limitations are ultimately linked to the infrastructure's inherited positionality: the cultural history we have already come to understand from chapter 1. This will show just how partial the disability access-knowledge of the infrastructure is and that the formation of categories it has developed is not, as it may present itself, all-knowing and innocent.

The step-free version of the Transport for London map of what is popularly called the London Tube (which includes the Underground, Overground, Docklands Light Railway, Thameslink, and the few remaining trams of south London) offers very specific strategic information about each of the stations, including the size of the step between the platform and the train, the size of the gap between the platform and the train, which stations have manual boarding ramps, which have "level access" boarding points, which have accessible toilets onsite, which have elevators, and more.[12]

The map can allegedly be used to plan one's journey; to do so, one would simply "check [one's] starting and destination station, plus any connections" and be able to trace out a "step-free" route. One should also remember to check the return journey, in case it requires an alternative entrance; additionally, as the map reminds its users, step-free changes between lines are sometimes limited to one direction. And one should also check for planned engineering works on the Transport for London website. Finally, one should also remember that "staff help may not be available at Thameslink stations at all times." A hypothetical journey that begins in east London, say, at Dalston Junction, and ends at central London's Euston Square leaves us the following trail.[13] We begin at Dalston Junction, a station marked as accessible, where the wheelchair user has level access to the train. From there, we alight at Highbury and Islington to transfer to the Victoria line. Upon arrival, the wheelchair user needs a manual boarding ramp to disembark. The map indicates that there is an internal interchange between the Overground and the Victoria line. The next transfer is at King's Cross St. Pancras, where we change to the Neapolitan lines. There should be an internal interchange between all lines and level access boarding points at all lines. From here, we finally journey to Euston Square, where we find that there is step-free access only on the westbound branches, meaning the return journey must follow a different route.

Using this map as a guide, the journey between the selected stations via this route is feasible, per the categories of access established by Transport for London. The infrastructure is offering here its own access-knowledge and solution for navigating transit as a wheelchair user; one need only follow these categories of access to make a successful journey. The categories are articulated straightforwardly: this station is step-free from street all the way to the platform, and then one needs a manual boarding ramp (deployed by a staff member) to board the train. This other station is step-free from the street and offers level access boarding all the way into the train. Or, this other station is inaccessible, or accessible in part. There are clear levels *to* access, in the infrastructure's conceptions, but they hold minimal nuance, as the infrastructure understands *access* as something that is singularly resolved by the provision of a ramp. Once the ramp is present, any barriers seem, to the infrastructure, to be magically resolved. Or, as TfL's website says, "At some stations staff will deploy manual boarding ramps to help you get on and off." The ramp becomes, through its mere presence and existence, the provision of access. In this, the ramp becomes an *access icon*: a symbol through which an infrastructure becomes satisfied with its own provisions of access because it is functioning under the assumption that these access icons are themselves access.

The wheelchair priority area is another access icon and another example where the infrastructure's access-knowledge is limited through an understanding that the mere provision of the priority area equates to access writ large. The Public Service Vehicles Accessibility Regulations of 2000 (PSVAR), a statutory instrument, stipulates that public vehicles must have at least one wheelchair space as well as the size of said space (including a reminder that it ought to fit both an assistive device and its user).[14] These regulations have shifted over the years, showing that the infrastructure's own access-knowledge is not static but rather constantly permutating—including, in the context of the rail regulations, to finally acknowledging that wheelchair users cannot be merely plonked into that space but must, in fact, maneuver to it, as other passengers not using wheelchairs might.

The regulations' detailed stipulations have resulted in various configurations of the wheelchair priority area on models of buses that are used throughout London's transportation system. We saw an example of this space in chapter 2. The type of disability access-knowledge offered in the regulations and, by extension, embedded and materialized in London's buses seems absolute, not partial. But here, again, the understanding of access is limited. The wheelchair priority area becomes another access icon—its provision is taken by the infrastructure as a signal that, in and of itself, accessibility is a given on buses. Indeed, as TfL articulates on its website in the section titled "Wheelchair access and avoiding stairs," "All our bus routes are served by low-floor vehicles, with a dedicated space for one wheelchair user and an access ramp. Buses can also be lowered to reduce the step-up from the pavement." Wheelchair access is thus reduced solely to the provision of these access icons.

In this sense, the level of detail given by PSVAR on a wheelchair-accessible area and the amount of information offered by the step-free Tube map would seem to be total, offered as axiomatic truths. There *is* a wheelchair priority area. Despite the number of inaccessible stations, there *are* some with ramps. These access icons are mobilized as materializations of the infrastructure's disability access-knowledge, and one should be able to parse the information and establish how one would navigate the infrastructure. The situatedness of this flattened access-knowledge of the system, however, comes into stark contrast with the complex crip access-knowledges of wheelchair users as they move through the infrastructure.

BELLIGERENT TECHNE IN ACTION

I offered the concept of belligerent techne in the previous chapter, briefly defining the term as the various ways wheelchair users interact with the infrastructure that go against its prescribed or expected use. In this section, I refine the definition and show examples of how a techne emerges when there is a gap or distance between experiential knowledge of how access truly works and an infrastructure's flattened and simplified understanding of access through access icons. Indeed, often, belligerent techne emerges in explicit response to bad forms of disability access-knowledge; in other words, uses of belligerent techne are deployed when there is misalignment between the legitimized and materialized knowledge of the infrastructure's makers and the realities of wheelchair users' own situated crip access-knowledges. I stated repeatedly that crip access-knowledges are complex and nuanced, partly due to the distributed nature of their acquisition and sharing. It is for this reason that wheelchair passengers, and other disabled passengers, are not duped by access icons. Users who understand and know the complexities of access see what is lacking in the infrastructure's material substrate. They very literally know it won't work and must find ways to fix, expand, hack, and better explore the context.

A *techne*, as Henry Staten reminds us, is "a practical knowledge that before it migrates to an individual mind-body has been accumulated across cultures over generations, centuries, millennia" (2019, 6). A techne is not a singular possession that an individual deploys by themselves but rather a system of knowing-making that inscribes behavioral and affective grooves that, over time, one becomes accustomed to following. As a techne leaves its trace, other wielders of techne must trace their own paths, loosely following these grooves but having to hone their own techne in tracing the grooves left before them. In this sense, Staten affirms, a techne is not merely the product of a practice but rather the *practices themselves*, which, "far from being unitary and self-identical, are decentered, distributed, perpetually adapting and changing forms" (2019, 24). A techne, he argues, is what ultimately enables human agency.

I use this understanding of techne to propose the idea of a belligerent techne, developed and deployed by wheelchair passengers of the infrastructure. In this sense, this type of techne goes against the grooves that we have inherited from the techne, or knowledge practices, of the infrastructure itself. The belligerent techne of wheelchair users rejects the grooves inherited by the system to create new grooves, in ways we saw in chapter 2. Belligerent techne therefore must emerge in the gap between the limited disability access-knowledge of the infrastructure and the much more nuanced and complex crip access-knowledges of wheelchair users. They must be belligerent because the alternatives that are offered as legitimate in the infrastructure are unsatisfactory. Nonbelligerent techne would follow the infrastructure's stipulations, happily following its prescriptions and scripts. But legitimacy is not enough to ensure access and, indeed, is not enough for survival. Wheelchair users' belligerent techne functions thus as direct responses to access icons—rejecting simplified retrofits in favor of more interdependent and, significantly, relational access practices within the infrastructure. In its belligerence, their techne becomes truant acts of freedom. I will illustrate.

I offered above two access icons that demonstrate the limited scope of public transportation's access-knowledge. Here, I will illustrate just how limited this knowledge is, and show how wheelchair users' belligerent techne are expertly deployed in response to them. Of interest here are also the many ways that wheelchair users' access-knowledges go beyond that of official sources, including Transport for London staff, because, as Kerstin put it, "they just don't get the system themselves." The first access icon was the ramp and the infrastructure's assumption that a map that shows the availability of ramps or step-free access at stations will provide appropriate and sufficient knowledge. As discussed, wheelchair users spoke about spending a significant amount of time trying to "work [out] the route" (Marie) to ensure that it is accessible. Aimee, for example, estimated that she spends "probably an hour a week" looking at the Transport for London route planner on their website. After telling me that knowledge is a trick, Anton described to me how difficult it is to determine the level of access available at stations despite the various categories developed by TfL:

> And there's a white wheelchair symbol, so that suggests great for wheelchair users. But no, there are no ramps at Upminster. So you can get there but you can't get off the train.

Others have had similar issues with the system's official designation of *accessible*, in particular when it came to the question of manual boarding ramps at particular stations, which require staff presence for deployment.[15] On the step-free Tube map, this type of station is marked with an *R*. They are among the most common accessible station designations on every line, with the exception of the Docklands Light Railway and the trams. At these stations, a wheelchair user needs to request that an employee accompany them to the platform, unlock the ramp from its holder, and be ready to deploy it. The staff member should then also ask what station they are heading to—if that one also requires a manual boarding ramp, the first employee must call ahead to request that another employee be present at the arrival station to help the wheelchair user disembark. Carl related experiences where the aspect needed for access—the employee with a manual boarding ramp—wasn't present at his arrival station:

> I had an experience where the guy with the ramp wasn't there and I attempted to disembark the train and the front caster wheel got stuck in the gap. —*Carl*

Previously, we also heard from Basil with a similar experience, where he resorted to being lifted out of the train by a group of football (soccer) fans. The category *accessible*, as stipulated by Transport for London on both the common map and the step-free guide, is therefore not a stable predictive category for the stations where access to the train requires a manual boarding ramp not only to get on the train but also to get off it. This can create friction with other stations, too, as was the case for Alan when he boarded a train using a manual boarding ramp and was heading to a station with level access only at designated boarding points (more broadly called *humps*):

> Assistance turned up with the ramp, put me in the wrong carriage. I asked him before we set off, because I know at Green Park you have to be in certain carriages to get the raised platforms [humps]. And I said to him, "Are you sure this is the right bit?" And he said, "Yeah, it's all right, this is where you need to be." Put me on the wrong carriage so we got to Green Park, which is where I thought I would just wheel off and it'd be easy, and there's a great big step, big gap.

If we return to our hypothetical trip from Dalston Junction to Euston Square and we know of these frictions, the accessibility question becomes much more complex than the map sets out. There is level access at Dalston Junction, which seems straightforward. However, we know that a manual boarding ramp is required upon arriving at Highbury and Islington to change to the Victoria line. This requires finding an employee at the departure station—Dalston Junction—before boarding the train to request that they radio ahead to Highbury and Islington to ensure someone is there with the manual boarding ramp. I hesitate to extend the scenario to King's Cross St. Pancras, as that station is categorized as both requiring a manual boarding ramp *and* having level access humps, depending on which of the lines one is using (and those differences are not indicated on the map!). What seemed like a straightforward trip (by London standards) on paper already begins to spin out with frictions.

Wheelchair users fill these gaps with belligerent techne catered to their own abilities. Rather than using the maps developed by Transport for London, the use of a "mental travel map" (as D put it to me) is preferable. These mental maps are developed through their own crip access-knowledge. In this way, wheelchair passengers learn not to depend solely on the categories of access provided by the "legitimate" maps. They forge alternative categories of their own, such as "safe routes"—a term used by multiple interviewees. Indeed, the knowledge carried by these users is such that our conversations were filled with references to specific stations and how the interviewee's knowledge of them clashed with employees'.

> Mile End is officially not accessible, but you can change there. You need assistance from staff, they must call ahead, they have manual boarding ramps, it's not a problem. Whenever I say at King's Cross, for example, could you please call Mile End because I want to change there, they say, "No, Mile End is not accessible." That is right, but I don't want to leave at Mile End, I just want to change to another train; because they are on the same platform, it's not a problem. I don't want to get to street level. —*Kerstin*

Kerstin goes against a representative of the infrastructure: she tells staff members they are wrong, clarifies what she wants and why it is feasible, and, through her belligerence, carves out access. Her case is illustrative also of how the different categories of accessibility established by Transport for London itself are confusing and partial. Mile End is, truly, not marked as an accessible station on the most commonly found Tube maps (in other words, there is no wheelchair icon at that station). It is, however, marked on the step-free Tube guide as a station where one can switch between lines. Kerstin did not specifically mention how she learned that she could change lines there, but the fact that *she* knows this, but employees do not (or at least do not always), is an apt illustration of the complexity and depth of her access-knowledge. Kerstin has a mental mapping of where she can and cannot go. Other interlocutors, such as Aimee, take their mapping one step further:

> You just have to think differently about how you do things, so I've actually got a Tube map where I've just taken a black felt pen and crossed out all the stations I can't use because even if you've got the logos on the Tube map, it's just easier if you see ones that you can use.

Unsatisfied by the way the map currently describes and shows accessibility, Aimee finds it simpler to mark it up on her own terms, clarifying paths of possibility. Aimee hacks the original map and its legitimized categories and imposes her own categories and needs onto it. She turns the map into a much more friendly tool, making it more useful to her than it was in its original state.

Wheelchair users have developed other forms of belligerent techne based on their access-knowledges. These includes the collections of useful items, tools, and other equipment they carry to close the gap between their complex understanding of accessibility and the limited provisions of the infrastructure. These "tool kits," as Alan calls them, enable quick fixes in the process of navigating the network. Consisting of various items—such as a T-key to unlock the manual boarding ramps from their holders if a staff member isn't there, or even portable boarding ramps—these tools can go as far as creating new possible routes that are not officially recognized as accessible by the Transport for London map and TfL employees. Alan and Yvonne's use of a personal portable boarding ramp is an example of this form of belligerent techne. Starting at Richmond station in west London, Yvonne deftly picked up the portable lightweight ramp from the back of Alan's wheelchair. Richmond is an accessible station, with the use of a manual boarding ramp, but they didn't ask the staff members for help as we were changing over at Hammersmith to the Piccadilly line to head to Green Park station. Yvonne again deployed the ramp at Hammersmith station for us to disembark. There, right across the platform from us, was the Piccadilly line. Alan explained that he preferred this self-developed exchange at Hammersmith rather than continuing to Earl's Court. He knows that an exchange at Earl's Court would mean having to switch platforms (and therefore dealing with elevators and tunnels) rather than traversing a platform to catch the next train.[16] Importantly, Alan knows and articulates that what he's doing is against the prescriptions of the system. He deliberately does not ask for the ramp at Richmond station because doing so would require informing staff that he would be alighting at Hammersmith—something the staff would not recognize as possible because of their own limited knowledge of the system (i.e., the map does not signal that this is a possibility). They would likely spend time questioning him, further delaying a trip that, on his own terms and based on his own access-knowledge, is perfectly reasonable.

It is clear that in the face of legitimized access-knowledge inscribed in the official tools and employees of the infrastructure, wheelchair users' belligerent techne goes against the grain. Through developing tool kits and customized maps and mobilizing their complex access-knowledges—replete with categories of their own making—they weave through the network in ways that are often in opposition to what the designed materiality of the system would require.

We continue to see clashes and gaps between access-knowledges in the case of the bus wheelchair priority area described above. All TfL buses have a single wheelchair priority area, following the stipulations of the Public Service Vehicle Accessibility Regulations.[17] As we have seen, this redesign of buses signals a shift in the infrastructure's designed materiality toward the inclusion of disabled people in public transportation. While it caters to the access needs of wheelchair users, this space has created new tensions on board. I detail here a couple of the types of tensions between knowledges brought about by the design of this area, and how wheelchair users solve them.

First, wheelchair users are required, by the design of the space and vehicle regulations, to travel facing the rear of the bus. The wheelchair space was designed, it can be argued, in such a way that enabled the greatest independence for the wheelchair user: when traveling backward, wheelchair users do not have to be strapped in or secured by straps or belts.[18] This both enables the wheelchair user to independently maneuver into the space and takes away a task that might fall to the driver—securing the passenger to the area. The prescribed space has a stipulated minimum area per the PSVAR, and occupancy is established in the Big Red Book, the bus driver's manual. As the manual states, "You must make sure they have their back to the backrest and their brakes are applied or motor disengaged if they have one" (TfL 2014c, 69). Multiple wheelchair users have argued, however, that the space is not large enough, that the design is "not the most innovative" (Kerstin), or that it is difficult to maneuver into the space due to the handrails. Alanni speaks to that:

> You have to go backwards according to the law, so the pole makes it really hard because it's there and you're trying to go into this space here.

Alanni further described depending on their partner to lift the back of their wheelchair to put them into the prescribed position. For others, the position itself is a problem:

> I suffer sometimes with travel sickness and so I find it easier sometimes to sit the opposite way to how you're supposed to sit in the wheelchair space.
> —*Michael J.*

Michael's belligerent techne is to roundly ignore the prescriptions of how to occupy the wheelchair space. As he told me he did this, he paused for a second and, with a slight chuckle, added:

Yes, and I fully admit and I probably shouldn't admit that but I do. . . . I'm not five, I do understand, I do know what the rules are.

Michael willfully rejects the rules of the infrastructure inscribed in the designed space. He opts, instead, to make the experience of traveling more pleasant, positioning himself so that he faces the front.

Wheelchair passengers also described processes of negotiating access to the space with other passengers on the bus. Perhaps the most common issue that wheelchair users mentioned in navigating London transportation is the case of the wheelchair priority area debates. The Doug Paulley case, or the "wheelchair versus pushchair case" as it has also been called, has become the exemplar of this issue. Though the case occurred outside of London, in Yorkshire, it gained significant media coverage and is relevant to our discussion here.

In 2012 Doug Paulley, a wheelchair user in Yorkshire, attempted to board a bus. As he did, however, he came to realize that the wheelchair priority area was occupied by a person with a child in a pushchair. The bus driver did not ask nor require the occupant with the child to fold or relocate their stroller to make space for Paulley, who subsequently sued the service provider, FirstGroup PLC, for unlawful discrimination due to his disability. The case made its way through the UK legal system over a half-decade, eventually making it to the British Supreme Court, with a final ruling in 2017.[19] Paulley's situation was not unique; many wheelchair passengers I interviewed had similar stories to tell about struggling to board a bus due to the wheelchair priority area (as it is named by Transport for London itself) being occupied by a non-wheelchair-user. These instances have been baptized by one informant as "the buggy war" (Kerstin) and represent a "complex issue" (Anton) that may call for various belligerent techne depending on the wheelchair user's evaluation of the situation.

This evaluation includes, for example, whether there is a sleeping baby in the carriage, how many pushchairs are currently in the space (one? three?), and even whether the bus driver will enable the negotiation process or not. In some cases, interviewees informed me

it's the bus driver who doesn't even give me the chance to negotiate with the parent in the space, they just say, "No, there's somebody in the space. You can't get on, you'll have to take the next one." My favorite phrase. —*Sophie*

There often is, even according to the bus driver's manual, enough space for a wheelchair and an unfolded pushchair to share the wheelchair priority area, though there, too, the manual points out that the space ought to be cleared for wheelchair users who need it.[20] Even in cases where this message is made clear to other passengers, with the driver engaging directly, still the negotiation can be fraught. A case described by Sal is particularly illustrative:

> But this particular incidence [incident], it was one of the new Routemasters, so it had a driver and a conductor and I think it was virtually empty.[21] There were, what, five or six people downstairs? And the driver put down the ramp and as I started to go up the ramp, I saw that there was a man, presumed father, with a buggy with the wife holding the baby on her lap on a seat, but he refused to move the buggy. He just would not move the buggy. And the conductor came up and remonstrated with him, but he just wouldn't move. And eventually, the conductor said, "I can't force him to move, so you know, you'll have to get off. I'm sorry, I can't take you anywhere else." And I had even tried to explain that if the guy moved out of the way and I parked myself, there was room for him to put the buggy back in front of me and we would both be fine, but he would not move.

Driver involvement can help at times, but there is little else that can be done. The driver cannot, for example, require the other passenger to disembark the bus. Here again we see wheelchair users applying a variety of belligerent techne to handle these situations with varying degrees of success.[22] These might range from personal abilities and skills with the wheelchair (deftly maneuvering into the space to show the buggy user where they can fit, or even dropping in front of the bus to stop the bus from leaving until they've been allowed to board) to keen emotional and social negotiation:

> I put on my best little-lady-in-the-wheelchair face and ask really nicely. They'll either move really quickly and be really nice, over-making-up for it, or they'll do it really begrudgingly. And I'm sorry if they're not happy, I just think, "Tough." —*Jo*

To negotiate the space on the bus with other passengers, Jo "puts on" a particular persona and voice that she believes will incite people to react compassionately toward her and to collaborate in making space. That this is something that is needed for her to gain access to a space she technically has priority to might be contentious, but the point here is not to argue about priority but rather to show the layers of knowledge (of oneself, of the social situation, of the infrastructure's points of friction) that are mobilized by wheelchair users to do access work.[23]

In our conversation, Chiara also spoke of acting more disabled, more frail, as a performance. In their words:

> When I have to correct my knee, I have to stretch my leg straight out and then people start looking at you funny because you're stretching your legs out and you're in a wheelchair and what is this? I feel like there is an onus on us—and this is something that I've observed from other wheelchair users as well—there is an onus on us to perform disability, so we need to make sure that we're not getting out of our wheelchairs, that we're not moving our legs in ways that would suggest that we're not disabled.

In *performing* their disability, enhancing it, as a personal attribute, wheelchair passengers must negotiate two roles in their interactions with nondisabled people in transportation. The optics of being a disabled person who is independent (a marker of importance in a society of compulsory able-bodiedness, where productivity and independence are valued) is placed in contrast with being a disabled person who may, in fact, need assistance. Anton expressed this internal negotiation thusly:

> I've had people try and push me into shops if there's a ramp. In my manual wheelchair, if there's a little step like that, I can get over it. And I like to win that battle. I don't like people coming in and saying, "Hey! Can I push you in?" So . . . it's very double-edged. Because it's a very good piece of . . . they have a good heart, and they mean well, but I really don't want that. It's intrusive and patronizing. I like it, I like the fact that most strangers really want to help. I was struggling up a dreadful slope on Saturday and just rolled backwards and couldn't do it, and I spun around and I was out of control, and the nearest man just said, "Hey, let me give you push." I just needed it at that moment and it was great, so he got me out of trouble.

It is in the moments when wheelchair users need to enlist other passengers to help them—or, at the very least, not hinder them—that the performance of disability becomes useful. This is particularly true in cases where wheelchair passengers feel the need to cede to social expectations of disability, in large part due to the social narratives that disabled persons find themselves stuck in: seen as paragons of superhumanity and inspiration on the one hand, or scroungers faking their disabilities to take advantage of government benefits on the other. Wheelchair users told me of feeling policed in public transit or simply when moving around town, being confronted by "funny looks" (Alanni) or judgmental looks that seem to ask, "Oh, you can walk, then why do you need the wheelchair?" (Leda).

Thus we see that, through the development of alternative—belligerent—techne, wheelchair users forge paths of access for themselves throughout the system. It is in the ways this belligerent techne stands in opposition to the techne that has already worn grooves into the designed materiality that this techne gains its name, for wheelchair users are going against the grain of the system. The belligerence here is key, as it reminds us again that the crip access-knowledges of wheelchair users are more complexly woven and finessed than that of the transportation infrastructure itself. The issue that remains, then, is the question of legitimacy of these knowledges within the system.

BELLIGERENT TECHNE AND LEGITIMATE KNOWLEDGES

I have argued that different forms of access-knowledges exist in the context of public transportation infrastructure in London: the complex and nuanced crip access-knowledges of wheelchair users and the limited, regulated disability access-knowledge of the infrastructure itself and its core community of practice. Through the complexity of their access-knowledges, wheelchair users have developed diverse methods of belligerent techne to close the gap between what the infrastructure offers as access icons and their own understanding that accessibility is more than the presence of a ramp or a wheelchair space. Here, I turn to Alan to begin offering some concluding remarks on the work of belligerent techne in the context of the diverse knowledges present in infrastructures:

> That is one of the things that as a disabled traveler you have to do. They're doing their health and safety assessments, but you have to do your own, don't trust what you're told. . . . One of the adages we developed early on, wasn't it, "They're in charge, but you're in control." It's their station, they know the rules, they're the ones that can say yes or no, but they can't make you do anything you don't want to do. So you can take as much time as you want, or as you need, because they can't hurry you up.

Alan gives us a crystal-clear articulation of what he sees as different priorities, different schemes of knowledge, being put to him: "their" health and safety assessments—meaning Transport for London and their associated transport providers—versus "your own" assessment of the situation—the disabled passenger's. It is the latter that he places in the realm of knowledge that is applicable to him, of what can be trusted as accurate in terms of how to interpret situations within the infrastructure. It's clear to Alan that the infrastructure, the disability access-knowledge it offers, is not superior to his own and how he orders his life. Rather, he trusts what he knows, the crip access-knowledge he has accrued over the years, and, importantly, has gained from and shared with others.

STS analyses of knowledge have often returned to the 1980s debate (continued through to the early years of the twenty-first century) on types of expertise. The question of lay versus expert knowledge might offer us a framework for considering the infrastructure's knowledge as expert knowledge, developed by engineers, urban planners, architects, and other authorities, and placing it in direct opposition to the lay knowledge of disabled passengers.[24] This mapping might be relatively easy to accomplish if we allow, for example, that standardization is undertaken by "experts" and agree with Wynne that lay knowledge is that which "allows control [of] a contextually dense and multidimensional reality in which adaptive flexibility towards the uncontrolled is still recognised as a necessary attribute" (1996, 70). What I find ultimately unsatisfying about this mapping, however, is that it does not allow for much discussion beyond the initial categorizations of lay versus expert knowledges. Falling back on those categories firmly ensconces knowledges as one or the other. Here, an analysis that uses the categories of crip and disabled access-knowledges allow for a more layered discussion of what knowledges exist, how they are differentially mobilized, and how they relate. In so doing, it shows various forms of knowledge as being more or less resistant to one another.

My analysis resists the bifurcation of defining some knowledges as expert and others as lay because it simply posits that disabled people "*are* experts and designers of everyday life" (Hamraie and Fritsch 2019, emphasis mine). Rather than reifying categories of expert and lay, the crip feminist lens enables us to understand the knowledges of the system and of its disabled users as functioning within different forms of knowledge, each one dependent on their own bounded and situated positions. We have in the infrastructural regime what can be interpreted as disability access-knowledge, which sits in contrast with the crip access-knowledge that is deployed in wheelchair users' belligerent techne as they interact with and, ultimately, shape the infrastructure to better suit their own understanding of accessibility.

Disability access-knowledge might mobilize a traditional understanding of expert versus lay, where experts are those who provide the solution and the disabled lay person is the one who may know what they need but not necessarily how to provide a solution. We can see how the infrastructure offers itself as the expert in presenting solutions: it provides the step-free Tube guide; it follows prescribed accessibility regulations; it provides wheelchair priority areas and manual boarding ramps that enable access; the bus driver's manual dictates priority for wheelchair passengers . . . The infrastructure gives what it understands as solutions for access issues, literally mapping the most efficient way for a wheelchair passenger to navigate it. These solutions are continually materialized in various technologies such as the provision of the ramps, the bright blue and white wheelchair icon painted on the floor, the loud siren that alerts other passengers that the telescoping mechanical ramp is being deployed on the bus. These attempts at access are limited to a specific understanding of access as being the movement, and even assimilation, into the infrastructure. It is insufficient, however, because it essentializes access as a legal requirement fulfilled through reasonable accommodations. More specifically, in the infrastructural context, it further reduces these accommodations to categories ("accessible with a manual boarding ramp," "step-free access," "wheelchair priority area") and standards such as the PSVAR's index of requirements for wheelchair access.

As we now know, having taken the time to explore the belligerent techne that wheelchair users developed in response to simplistic access icons, these essentialized definitions of accessibility are too limited when faced with the experiential, lived, crip access-knowledges of wheelchair passengers. A wheelchair priority area does not accessibility make; a step-free Tube guide does not provide all the answers (and it certainly does not ensure the presence of staff members with manual boarding ramps). The various links between access provisions that need to add up to make an accessible journey are, more often than not, dependent on the wheelchair passenger's know-how to yoke them together.

We see, in the varying interventions in response to the infrastructure's knowledge, wheelchair passengers manipulating the limits of convention to create true access for themselves. In this sense, the theory of access that I offer here is one that is deeply relational and interdependent, contingent on the mobilization of a range of skills and capacities, shared and distributed social networks, and active manipulation of the infrastructure's materiality. In the case of the wheelchair-accessibility maps, wheelchair users' access-knowledges enable them to forge better accessible paths than the step-free or common Tube maps. Beyond that, wheelchair users intervene with techne of their own that further create access—developing their own maps or carrying tools that allow them to bypass the infrastructural categories of access. In the example of the wheelchair priority area, in defying the rules of transport and riding "backward"—which is to say, facing forward—to enable their own comfort while traveling or deploying social skills to negotiate access to the space, wheelchair passengers become belligerent users. In so doing, they challenge the infrastructure's historical communities of practice (bus drivers, normate users) to inscribe their existence into the infrastructure itself.[25]

The belligerence toward the dictums of the infrastructure's designed materiality, pulling on the values of interdependence and relationality, is what creates access. Ultimately, crip knowledge and "crip science" (Piepzna-Samarasinha 2018) bypass the original infrastructural categories and standards, such as *accessible*, *step-free*, or *not accessible*. Instead, crip access-knowledges construct labels of their own: *areas I can get to*, *safe routes*, *requires personal ramp*, *call TfA for help*, *tell TfL they messed up*. In navigating public transport, wheelchair users are not taking for granted the status of the infrastructure as a controlled environment whose categories they need to follow. Rather, through experience (and sharing of experience), information about the shape and characteristics of the network, and management of their bodies (both biological and extended, i.e., the wheelchair), they develop mechanisms of engagement with an otherwise unstable and unpredictable environment.

The central values of crip access-knowledges of interdependence and relationality are significant to carry forward. This stands in stark contrast with the infrastructure's disability access-knowledge, which, in its choice of rear-facing wheelchair spaces, has positioned independence as a core value. Interdependence, as Hamraie and Fritsch assert, "acts as a political technology for materializing better worlds," offering spaces of alliance and solidarity (2019, 13). We see interdependence in Alan's travels with his partner, where she anticipates and helps with the deployment of the ramp. We see it in Jo's approach of the "little-old-lady-voice," where her understanding of access is one that means she needs to convince other passengers on the bus to see her needs as on par with theirs and move aside so she can board. These examples of belligerent techne become more powerful when we understand how they have also been entangled in a wide web of crip access-knowledges in their distribution through diverse networks (such as Transport for All).

I offered here ways in which diverse regimes of access-knowledge are present in transportation infrastructure—crip access-knowledge and disability access-knowledge. I highlighted how infrastructural access-knowledge concretizes definitions of expertise in ways that are materialized in the infrastructure through diverse methods (e.g., spatial design, way-finding maps). From there, we saw how these concretizations and categories are often at odds with the daily experiences of wheelchair users, who contend with these misplaced concretizations and build their own forms of access through belligerent techne. My goal here is to show how disability access-knowledge flattens its solutions to access to the provision of access icons, an attempt at standardization that seeks to control the much more complex, relational, and interdependent nature of accessibility. This is something that disabled people's access-knowledges are already aware of and that their techne reflects.

In the case of transport infrastructure, concretized and reified disability access-knowledge in the guise of access icons has resulted in a rigid materiality into which social changes struggle to embed themselves. As a result, wheelchair users' experiential knowledge of themselves and of the network provides them with an alternative way of interacting with the infrastructure. As wheelchair users create networks of crip access-knowledge, of which Transport for All is but one node, they can distribute their own expertise and collect that of others. The question of legitimacy, however, remains: the legitimacy of wheelchair users belonging in this infrastructure, or even the legitimacy of their claim to the wheelchair priority area. In this sense, how legitimate is the knowledge of wheelchair users, and how legitimate in the eyes of the network are their claims that the infrastructure simply does not work well for them? The work of legitimatizing user experiences as forms of concern that are visible and legible to the infrastructure is partially done through Transport for All. However, as we will see in chapter 4, infrastructure studies as a field has struggled to fully grasp that user experiences and interactions within systems ultimately shape these systems. This is largely due to the persistent metaphor of *invisibility* that permeates work in infrastructure studies. This metaphor is one that easily sidesteps questions of legitimacy, belonging, and relationality. In the next chapter, I work to untangle the invisibility metaphor and ask that we push infrastructure studies farther.

FOUR / Beyond Invisibility / *"I feel a lot of the time that I'm invisible still."*

Michael J. and I never met in person. I interviewed Michael about halfway through my field research over Skype. That day, our internet connection struggled to stabilize, and we kept out cameras off to help. Despite this, the connection dropped multiple times, but Michael, a man in his thirties who volunteers with local organizations, generously persisted in conversation with me. About ten minutes into the interview, the connection dropped for the second time. "Are we back?" he asked, to ensure we were both back online. He laughed, "Yeah, I was going to suggest meeting for a coffee, but I wasn't sure you'd know where Hounslow was." We laughed and picked up the conversation, which shifted dramatically in tone as I repeated the question I had asked before the interruption: Tell me, what are your general experiences with transportation like—more positive, or more negative?

> As I said, in the past few years, I have noticed positive changes with regards to the Underground and London buses and docklands and everything, but if you're asking me do I find it a positive experience, no, I do not find it a pleasurable experience, I do not find it a positive experience, because I feel a lot of the time that I'm invisible still, you know?

It wasn't originally a line that stuck with me. I had heard similar sentiments expressed before in various ways. By then, I had heard wheelchair users tell me they often feel ignored, pushed aside, or passed over, so it wasn't a surprise to hear Michael say he felt invisible. However, as I sat down to analyze the interviews, I found that Faith, a Northern Irish woman in her late thirties and a transplant to London, had also used the word *invisible* in our interview:

> For as long as disabled people are invisible, largely, people aren't going to change their attitudes because they're not going to be confronted with them.

The question of invisibility is nearly unavoidable when working in the field of infrastructure studies. However, it is not usually deployed in the way that both

Faith and Michael do here. Michael and Faith are speaking of disabled people as invisible—overlooked, ignored by society. They articulate invisibility as a negative experience, associated with the inherited ableism of society at large, in ways that impact their daily lives significantly—including on public transportation. In infrastructure studies, however, *invisible* is a descriptor applied to infrastructures themselves. Numerous are the scholars in the field who speak to the invisible nature of infrastructures, leaning into an often-cited definition of infrastructures as "designed to become invisible as [they are] stabilized" (Lampland and Star 2009, 207), only becoming visible again upon breakdown (Edwards et al. 2009; Graham 2010; Edwards et al. 2007; Lampland and Star 2009). In this sense, invisibility stands in for mundanity—that which sinks into the background of our daily lives and, in so doing, becomes unremarkable, unquestioned, easily manipulated and used. Thus, it is when infrastructures break, or cease to operate, that they become visible as the pieces need to be put together again.

Due to the question of maintenance, the amount of work that goes into keeping infrastructures supposedly invisible, some scholars have pushed back against the metaphor of invisibility in infrastructures. Given the need for constant repair and maintenance work, infrastructures are, these scholars argue, mundanely visible (Harvey and Knox 2012; Larkin 2013, 2008; Ureta 2014). The debate on how and when to use the metaphor of invisibility has been brought to bear even on the Global North–Global South axis, where some argue that the disrepair of infrastructures in the Global South makes them particularly visible (Furlong 2014; Graham 2010; McFarlane 2008)—or they are hypervisible in instances of monumentality, as public displays of the "spectacle of technological promise" (Schwenkel 2018).

Despite the debates surrounding the invisible nature of infrastructures, Star's quote that infrastructures are "by definition invisible" has become so common as to be almost invisible itself in the sense that it has become largely accepted.[1] Some authors even deploy it as a primary characteristic that infrastructures must fulfill—"any genuine infrastructure is mostly invisible" (Edwards et al. 2009, 370). But to whom is it mostly invisible? Much like previous scholarship on the norms of science (Merton 1973), it is not always obvious what these characteristics or norms are truly articulating. Is *invisibility* deployed in the descriptive or prescriptive sense? Is it an observable state or something infrastructures are tending toward? In some cases, invisibility seems to stand in as a description of how its community of practice experiences the infrastructure. In others, it seems, instead, that it ought to be understood as a goal toward which planners and maintainers of infrastructures should work. This vagueness means that its use in this scholarship is rarely satisfying.

In my conversations with wheelchair users, the word *invisibility* was never deployed to describe the public transportation network in London. Nor were other words that often stand in as its semantic equivalents, such as *mundane* or *unremarkable*. Rather, the words my interviewees used were always negative, connoting experiences that any of us might have with things that do not quite work as they might: *circuitous, challenging, frustrating, othering*. No one in infrastructure studies has claimed that, rather than being invisible until they break down, infrastructures are, in fact, challenging until they work. Except that, in ways that I hope to recover here, this is already present in much of the original articulations of invisibility in the scholarship of Leigh Star and coauthors. In the process of defining infrastructures, the authors begin with what they tellingly call a caveat: "We hold that infrastructure is a fundamentally relational concept" (Star and Ruhleder 1996, 113). This statement precedes any further specifications or definitions of the concept of infrastructure, of which visibility upon breakdown is one. Indeed, always sensitive to those at the margins, Star refines this idea of relationality, poetically stating, "One person's infrastructure is another's topic, or difficulty" (Star 1999, 380).[2]

Given the mismatch between theoretical descriptions of infrastructures and wheelchair users' experiential descriptions of them, I will use this chapter to focus on questions of the invisibility and relationality of infrastructures. As I have argued, the crip access-knowledge of disabled passengers of public transport enacted through their belligerent techne is a shaping force within infrastructures—users forge access and, in so doing, actively form the infrastructure. Their belligerent techne is predicated on interdependence and knowledge practices that ultimately show how infrastructures are not ever truly invisible. Here, I expand on the notion that infrastructures are first and foremost *relational*, and, depending on whether the material, social, and relational components align, they may become invisible or not. I argue that the invisibility metaphor in infrastructure studies has taken precedence over key articulations of relationality. This has often led analyses of infrastructures down a functionalist path that obfuscates questions of positionality in the network. These analyses, predicated on functionality and invisibility, can also too easily set aside how said functions may not be equally or equitably distributed. Thus, I argue that infrastructure studies ought to question the meaning and use of *invisible* as a metaphorical descriptor, understanding it as potentially flattening users' experiences of networks. Here, the crip feminist lens I use to analyze and describe infrastructures does not allow me to accept invisibility as a characteristic for an allegedly functional infrastructure. I instead investigate what is meant by invisibility and functionality for infrastructures—invisible and functional for whom? In so doing, I center situatedness and relationality to argue that infrastructural invisibility is a laudable goal that nonetheless occludes disparate and plural experiences.

To do so, I want to first establish, through interlocutors' narratives, how their experiences with infrastructure already contest the idea of infrastructural invisibility. The various barriers that wheelchair users face in navigating London's transport system (despite slowly improving conditions) and their own discussions of those barriers help illustrate how foregrounded infrastructures are for misfit users in the infrastructural context. Some examples will now be familiar, but rather than highlight wheelchair users' belligerent techne here, I highlight how the barriers to access create a cumulative access problem that renders London's transport network anything but invisible to these users. I explicate the condition of *misfitting*, feeling both invisible and hypervisible within the infrastructure, as that which enhances infrastructural visibility. I finally address the concept of infrastructural invisibility as related to the concept of readiness-to-hand, questioning whether how the characteristic has been used as a defining concept in infrastructure studies has homogenized experiential accounts. Ultimately, I argue that we must set aside the invisibility metaphor if we are committed to seeing diversities of experience and knowledges in infrastructures and to truly grasp the disparate consequences of an infrastructure's designed materiality.

ACCESS ICONS AND THE BARRIERS TO ACCESSIBILITY

Perhaps one of the slipperiest concerns with invisibility is that it can apply in our case study to more than one character. A telling of the infrastructural side of the story would warrant focusing on how public transportation in London enables well over one billion passenger journeys per year. As a result of this seemingly smooth integration into the very fabric of so many Londoners' and tourists' lives, one may be satisfied by the description of infrastructures as invisible. I argue, however, that this is too simple an application of the concept, especially when we consider infrastructures from the perspective of user experience. So, we can describe the infrastructure as invisible, if we follow the paths that infrastructure studies has forged. Or, listening to the experiences of wheelchair users, we might begin to see how the other character who may be described as invisible is the misfit user—the user whose access-knowledge the infrastructure's designed materiality does not include.

For Michael J., the use of the transportation network is what makes him feel invisible. The network itself is not described that way. Rather, he and every other interlocutor I spoke to were able to discuss various features and characteristics of London's transport system in very specific detail: where staff are generally friendly and supportive, their "safe" routes, their biggest areas of concern. We must explore what these feelings of invisibility and visibility mean in relation to transportation infrastructure. What do we learn about infrastructures when we problematize static definitions of the invisibility of infrastructure and offer, instead, in/visibility as a category related to experience? To follow through with a discussion of transportation infrastructure as invisible would amount to a poor understanding of the experiences of wheelchair passengers. As we have seen, the amount of work wheelchair users undertake to create access for themselves should make anyone now familiar with the case question whether infrastructure can truly be said to be invisible to them. These passengers' nuanced crip access-knowledges show their keen awareness of the various fault lines of the network, even where allegedly accessible features, or access icons, have been provided. We can complicate the discussion of invisibility by not only applying it as a definition of infrastructure itself, but asking, What is invisible to whom? When?

Michael J. grasped for the term *invisible* to describe how he feels in relation to the infrastructure, rather than to describe the infrastructure itself. Similar themes of invisibility emerged in interviews in various guises, often in reference to feeling ignored, bypassed, or left behind by the infrastructure. And, often, these discussions were framed by passengers' frustrations with an assortment of barriers that wheelchair users must navigate despite the new access provisions introduced in the past two decades. So, while it is important to have remarked on how the access landscape has, in some regards, improved and adapted in response to the shifting context around disability rights, we must also pause and show the gaps that wheelchair users brought up in interviews. First, this will illustrate some of the realities of navigating the system while disabled, enabling a deeper discussion of how this goes against received wisdoms of the invisibility of infrastructure.

While the narrative of steadily improving access provisions, particularly promoted around the 2012 Olympic Games in London, might be easy to repeat, it is also one that interlocutors challenged. In a conference room at his work, I asked Alex Lyons, "Did you feel a difference during the Games?" His response came quickly:

> Yeah, definitely, completely. Absolutely. Whether it stuck? No.

Barriers to access for wheelchair users are as diverse as wheelchair users themselves. I might spend time, as I have done in the past, parsing them into more clearly bounded categories and themes. I have, in the past, argued that it is useful to distinguish among the types of barriers that wheelchair users face, following from a conversation with Marie (Velho et al. 2016; Velho 2017). When I asked her about what issues she encounters while traveling around London, she responded, "Are we talking about the physical barriers to accessible transport? Are we talking about attitudinal barriers?" While I still believe that the heuristic categories of physical versus social (or attitudinal) barriers can stand up to some analytical scrutiny, I have now argued that all barriers emerge through an infrastructure's designed materiality, which effectively *is* the deeply entangled sociohistorical form inherited through shifting perspectives and social arrangements in the infrastructure. This is owed to a crip feminist analysis that integrates feminist new materialist concerns into the complex and often ironic characteristics of materiality. So, while I do offer some identifiable themes of barriers in the following paragraphs, I have opted to speak about them as often as I can through my interlocutors, who go back and forth among the many possible categories of problems.

The simple act of entering the infrastructure can be difficult for wheelchair passengers, despite the provision of access icons such as ramps, which, themselves, are not stable categories of access. Indeed, concerns with ramps were often cited, including cases where the ramps are broken:

> There's also been once or twice a couple of issues where a bus has been in service and the ramp hasn't been in working order, which I've been told should never really happen because if a bus ramp is out of service or broken, essentially, the bus shouldn't leave the depot. *—Carl*

Or, if the ramp isn't broken, the bus driver may ignore the wheelchair user and not deploy the ramp to allow them to board at all, whether because they did not see that a wheelchair user required it or due to purposefully ignoring their access needs:

> Also, I've had bus drivers just plain refuse to let me on. Back in . . . actually, I still have the notes for this. Back in . . . Back at the end of last year, I was in the Ealing area and I attempted to board, I believe it was a 207, and the bus driver . . . He knew I was there, I heard the buzzer go, I pressed it two or three times, he refused to let me on. He didn't attempt to deploy the ramp, he didn't talk to me, he just tried to drive off. —*Char Aznable*

Say that the wheelchair user comes across a bus that they are able to board. They may still face access concerns with the wheelchair priority area on the bus. I have discussed the various issues surrounding the "wheelchair priority area," including priority debates with caretakers and young children in prams. "Battle of wheels" quickly became the code I used whenever wheelchair users discussed what other interlocutors referred to as the "buggy war" (Kerstin). The "battle" and "war" metaphors are useful stand-ins to understand how tense the situation can become when a wheelchair passenger is boarding a bus and the so-called wheelchair priority area is occupied by a parent or guardian with a child in a stroller. In 2015, the disability charity Leonard Cheshire surveyed over 179 wheelchair users throughout the country and found that 92 percent have been refused access to a bus and 61 percent identified buggies as the "biggest problem they faced" (Leonard Cheshire 2015). As we have seen in the case of Doug Paulley in Yorkshire, the question of who has priority to the wheelchair space on a bus is so fraught that it became the topic of heated media coverage and legal action in the United Kingdom in the mid-2010s. It was apparent in interviews the extent to which this was the case for my interlocutors as well, as all but three interviewees discussed the "buggy war" to some degree. While in some cases it was a parent or guardian who was adamant that they would not move, in others, wheelchair users' frustration was directed to the bus drivers who wouldn't allow them to attempt a negotiation for the space. Often, interlocutors signaled that their frustrations about the battle for the bus space are linked to their knowledge of why those spaces exist in the first place. If, as I described in the first chapter, access aboard buses began in the mid-1990s after disability rights campaigning, the wheelchair priority space was born out of disabled activists' engagement with public transportation. For many wheelchair users, it signals key questions of choice and history that parents and guardians with pushchairs are not engaging with or not aware of.

> In the old days, when all buses were old Routemasters, you didn't take a buggy on the bus. So, you had a small pushchair and you would fold it and

> put it in the luggage hold, and take your baby like that. . . . Then wheelchair users started campaigning for a wheelchair space and the wheelchair spaces were implemented, and that dragged in its wake the possibility of bigger and bigger buggies to use that space. *—Anton*

This rationale was brought into our conversations many times, with wheelchair users describing their frustrations with buggy users (and, particularly, with bus drivers) as related to a general lack of disability awareness or training.

Readers have likely noticed that many of the barriers discussed thus far are related to buses rather than the Underground. As all London buses are (technically) accessible for wheelchair passengers, many do prefer navigating the capital on the bus network. When I asked Jo what advice she would give to wheelchair users just beginning to navigate London's public transportation system, she simply said, "Ignore the Underground and plan." It's not difficult to see why that might be her suggestion given that only about a third of Underground stations and half of Overground stations have some degree of access. Despite that, wheelchair users can and do use the stations that are available to them, but there, too, they encounter similar barriers.

The issues surrounding access to wheelchair priority areas become fewer and farther between in interlocutors' narratives of their experiences traveling on London's trains. This may be because each train has at least two wheelchair spaces and, in the case of many types of train vehicles, ample space for standing.[3] In my conversation with Diana, she identified the conflict between buggy users and wheelchairs users as being particularly evident on London buses due to there being a single space for both to "share":

> So, in London, unlike a lot of England, there is generally only one space, so that automatically creates conflict, although there are a number of buses where there's actually multiple spaces.

There are some bus designs in other cities in the United Kingdom and Europe that provide two spaces on board that can be occupied by both wheelchair and buggy users, likely reducing the tension between them. The same is true on London's trains. But even if the tensions around access to space are minimized, the barriers are still multiple. This is not only related to the general lack of step-free stations. Even stations that are supposedly step-free come with issues:

> It's still as if . . . it's still as stressful if not more stressful because the wonder of technology is the wonder that it ever works. *—Michael J.*

In our conversation, Michael was referring to a broad collection of technologies that can fail. The discussion led us to list not only the automatic ramps on London's buses, but also elevators in London's step-free stations and the manually deployed ramps at stations that require them. Elevators feel like an obvious point of friction for accessibility concerns—indeed, elevators are another access icon whose mere presence is taken automatically as constituting accessibility. However, they are another access icon that often breaks down, and the breakdown of an elevator at a station that requires them means that station is functionally inaccessible to anyone who depends on them. A broken elevator or escalator means facing stairs, which, at London's deepest stations, are feared by anyone with an ounce of self-preservation. Covent Garden, for example, is known for its 193 steps, with an audio announcement reminding passengers that it is the rough equivalent of a fifteen-storey building. Still, it is nowhere near the worst of the stations in London; Hampstead has 320 steps. As is perhaps to be ironically expected, neither of these stations is step-free, despite elevators enabling access for non-wheelchair-users. When elevators down to these stations fail, accessibility for many people beyond wheelchair users is compromised. Climbing up or down a fifteen-storey building is not something most people are willing to embark on.

Elevators and ramps can fail for a wide variety of reasons. Ramps, for example, such as the mechanical telescoping ones on buses, might seize up due to mechanical failure,[4] or incompatibility with the environment in which one is being deployed (i.e., street furniture that blocks its path or curbs that are too high):

> The ramp has a kind of sensor, so sometimes when the driver pulls out the ramp, the ramp goes back because it's not actually that there's a problem, it senses something in the way so it gets back. —*Um Hayaa*

Ramps at some Underground or Overground stations, however, are dependent on humans, and their failures are somewhat different. These foldable yellow and black ramps are found at stations that are considered step-free from street to platform and require a staff member to deploy them for wheelchair users when they arrive at the station. The staff member must then communicate with the passenger's destination station to inform them that a ramp is required there, too, if that station also requires the ramp (some stations are completely step-free thanks to boarding humps). Before boarding a train at a station, a wheelchair user has to find a staff member and notify them that the ramp is required.

> At the gate there's supposed to be somebody there and I'm supposed to say, hello, I'm traveling to, from, and I need some assistance, please, I need the ramp. Now that person may acknowledge you, maybe not, they may come down and help you with the ramp, or may radio somebody to come and assist you, but you don't know what's happened because there's been no communication. —*Adam*

While the wheelchair user boards the train, the staff member needs to communicate and ask at which stop they will alight. If that station also requires a boarding ramp, they must call the station and inform them. This moment is crucial: if a staff member is not present at their destination, a wheelchair user might be stuck on the train until it reaches a station where they can alight without help, or they might require help from other passengers to let staff know they are stuck. As a result, the thought of whether or not staff will be present is a common concern.

> Well, we always arrive into Euston on the train and that has got better over the years, but there's always the, "Will the man be there to get you off the train?" and that's a bit of . . . that's the first concern, are you going to get off the train? —*Jo*

Interviewees described cases where they had to disembark the train without the presence of staff members with the ramp. Carl is a young wheelchair user and described his skills on his wheelchair as "not poor," having played wheelchair basketball in the past. Despite his abilities, he says he has got himself stuck "a couple of times" when he tried to disembark on his own:

> I had an experience where the guy with the ramp wasn't there and I attempted to disembark the train and the front caster wheel got stuck in the gap and the main wheels were fine but the front one got wedged in and it was kind of a case of, you know, I hope the guy doesn't just drive off.

Coping with the many barriers to access in London's transportation system is thus not only a psychic burden of not knowing what to expect (e.g., will the buggy user make way, will the driver advocate for me, will the ramp work, will the elevator function) but also contains real possibilities of bodily harm and abuse. Interlocutors shared some harrowing experiences that, though not explicitly related to using transportation, can generate further anxiety around the thought of merely leaving one's home.

> I was sat at traffic lights waiting to cross the road and I was pushed off my chair into oncoming traffic, and [someone] shouts behind, "Get out and run, you fucking lazy bastard." I'm desperate to get across, thankfully traffic stops. . . . But that's not the only form of verbal and physical abuse I've had. I've had a number of incidents. —*Adam*

In less extreme cases than Adam's, interviewees recounted feeling like they were being patronized or treated as objects of pity and charity.

> Some are a little bit patronizing. I've had one bus driver ask me, "Are you ok to travel alone?" I just kind of stared [at] him. It's just like, I am traveling alone. Of course, like, this feeds into the wider thing of society's perception of wheelchair users and what we must be capable of. —*Chiara*

Various wheelchair users described to me what they saw as a "shitty attitude" (Faith) toward disability, one that inevitably mired their interactions with nondisabled people in stigma. Negative interactions were seen as the result of one of two possible narratives: either the disabled person is a "benefit scrounger" faking their disability to gain government aid, or they are a legitimate disabled person who can only be the object of charity. We see here a continuing thread from disability history—the deserving and underserving poor of Elizabethan times lasting into the twenty-first century. Though this isn't the perspective of all people, some data points to 36 percent of the British public believing that disabled people are less productive than nondisabled people (Aiden and McCarthy 2014). These negative social perspectives from nondisabled people continue to shape and inform wheelchair passengers' experiences, including their ability to even board a bus, as we see in the debates of who does or does not deserve access to the priority area on the bus. Access icons cannot be confused for what disabled passengers know to be the nuanced nature of accessibility: slowly, deliberately crafted through negotiations, hacks, and sharing of knowledge. If anything, in the infrastructure's insistence that ramps or elevators themselves create access, it is missing out on how even these deceptively simple artifacts can be differentially designed and experienced. Alice Sheppard, disabled choreographer, dancer, and writer, coined the concept of "ramp joy," the feeling of absolute contentment and exhilaration when interfacing with a beautifully designed ramp. If we think of ramps as blunt tools of access rather than opportunities for joy and inclusion, we may never fully reach the liberatory potentials of designing *with* disability rather than *for* it.

In this sense, through the infrastructure's limited disability access-knowledge, evidenced by the variety of access obstacles listed above—and despite the access icons provided—it becomes increasingly clear that accessibility is a layered and relational issue: if the right conditions, right connections, right people are present and negotiated, including through disabled passengers' deployment of belligerent techne, accessibility may be forged. If not, however, accessibility falls apart and the infrastructure is painfully visible.

CUMULATIVE INACCESSIBILITY AND VISIBLE INFRASTRUCTURES

Naming accessibility as a *layered* issue is intended to capture how the concerns surrounding accessibility practices are not solely dependent on space (i.e., having the priority area) nor on non-stigmatizing behavior by other passengers (though that would certainly help). Accessibility is dependent on various concerns gelling to ensure that a journey does not require significant research (What routes can I take?), negotiations (Will the driver allow me in? Will there be space on board?), or fixing broken things (Will the ramp work?). As we have seen, wheelchair users must respond to diverse barriers with belligerent techne that do the work of forging access. However, what emerges in discussions with wheelchair users is that these barriers cannot be understood as isolated incidents that one might encounter on a journey. A pattern of accumulation became apparent in our conversations: a broken ramp, a person in the wheelchair priority space, a bad-humored driver, and so on weren't separate instances. The power of these barriers came, instead, from their cumulative nature: a series of seemingly small issues that together constitute a looming collective obstacle to wheelchair users' movement throughout the capital. This is clear in many interlocutors' descriptions of leaving plenty of "cushion" time in their journeys in order to deal with obstacles as they arise. Take the following four excerpts from different interviews:

> Again, there had been incidents when I have appointments and although I came out half an hour from my house and I know that going to that appointment would only take five minutes, I plan beforehand half an hour in advance. However, sometimes it happens that I wait for four or five buses because either pushchairs or the attitude of the driver or the ramp is broken or the driver himself doesn't know how to operate the ramp. So there is a full range of different obstacles. —*Um Hayaa*

> There's been numerous times, and I'm not exaggerating, where I've been sat at a bus stop waiting for a bus, one that I can get on because it's either full of people or the wheelchair space is being used by prams or pushchairs or whatever, and to find a ramp that's actually working can be very difficult, and so it's either . . . I either have problems because it's too crowded or it's basically not working, and there was one experience I had where I had to wait for seven buses to come before I could actually get one, before I found one that I could actually get on. Which was absolutely heartbreaking, because I was just sat there, you know, and each of the buses that I tried to get on the ramp either was broken so wouldn't come out, it would come out but then go straight back in again, or the bus had a wheelchair on already, which is fair enough, or prams, or it was too crowded, or whatever. It was one of the worst days in my life, to be honest, because by the end of it, I was just thinking, do you know what, stop the world, I want to get off! *—Michael J.*

> I once had to wait forty-five minutes at the bus stop because every single bus that came had two buggies in the wheelchair space and the driver was either unwilling to ask them to move or the passenger refused to move. What should've been an hour-and-a-half commute then becomes a two-and-a-half-hour commute because of that alone. *—Diana*

> Traveling by bus is very stressful from the moment you arrive at the bus stop . . . because you don't know what's going to happen. You don't know whether you're going to get on, you don't know whether the bus driver is going to ignore you, not even to deploy the ramp, which happened on Saturday. You don't know if there are going to be two buggies and whether if there's one buggy if it can be folded or if there can be an arrangement. It means I will set off two hours early. I allowed two hours to get here, and it took me two hours. *—Anton*

None of these interviewees knew one another, and they all live and work in different locations in London. They come from different ethnic backgrounds, have different gender expressions, and are of different age groups. Our interviews occurred in different places using different media. Despite this, their experience with the ways that various accessibility issues accumulate in a single journey to a point where the travel time doubles or even triples is one that was echoed in many other conversations, showing a high level of generalization to all wheelchair passengers in London beyond the sample interviewed. The barriers to London's transportation system create anxiety and frustration around the lack of predictability and their consecutive nature. If taken individually, not getting on a train or a bus immediately might not be a concern given that London is a place where either mode of transport goes by every few minutes.

> I'm not going to say it's not a problem, but it's less of a problem in London. . . . Most of our routes are a matter of minutes before the next one comes along. —*Robert*

Robert, a wheelchair user who works at a transport service provider in London, articulates a slightly defensive view of the situation. A white man in his late thirties or early forties, he acquired his impairment due to an accident. To him, the fact that buses come by every few minutes ought to at least help with some of the access issues, although it should be noted that the gap can be up to ten minutes on less common routes in London suburbs and sometimes even longer, especially at night. Having been interviewed at his place of work may have influenced his response, as he noted repeatedly in our conversation that it absolutely *is* a problem, if one that is somewhat alleviated through the provision of additional service. However, as other wheelchair users put it, if the barriers are multiple and often consecutive, five minutes can easily turn into twenty or thirty. The first bus has two buggy users in the space and no one moves; the second bus has a broken ramp; the third bus simply does not stop . . . Together, these barriers begin to feel insurmountable and, perhaps, even more frustrating when the infrastructure insists that accessibility provisions are there—in its access icons.

In the first chapter, I discussed the long history of transport in London. Despite the infrastructure's slow shift in conceptions of access, it seems that the experience of disabled passengers remains fraught. To an extent, the infrastructure has retained a stable image of the imagined passenger: a nondisabled commuter or pleasure seeker. As such, the service it provides has only marginally, and recently, embedded the needs of other passengers, which has resulted in public transport being experienced as a fractured infrastructure for its nonstandard users. These nonstandard users, with bodies other than the "standardized bodily package" (Moser and Law 1999; Star 1991), experience a materiality whose norms have not incorporated their bodily needs. As Star succinctly describes, "Part of the public stability of a standardized network often involves the private suffering of those who are not standard—who must use the standard network, but who are also non-members of the community of practice" (Star 1991, 43).

We see the private suffering of wheelchair users in their long waits, their various attempts to board buses, and the imperative to negotiate and justify their own needs as they interface with London's public transport system. We can also see it expressed in the consequent isolation, anger, and sadness that interlocutors expressed:

> It isolates you even more because your world is getting smaller, and smaller, and smaller, all the time, you're looking for more and more things that you can do as close to home as possible. . . . And then you end up being very isolated. —*Marie*

> I should be valued in the same way as any other customer and I just don't feel that we do that here. I think here we are a lesser form of person. That infuriates me. —*Faith*

As a result of this private suffering, wheelchair users experience a fractured infrastructure, which I define as an infrastructure that is experienced by its user as fragmented, broken, incomplete, or dysfunctional. To wheelchair passengers, London's public transportation is never in a fully functional state—it doesn't accomplish its primary task of enabling movement from one point to another in a steady state—and, because of this, the edges of the transportation system are utterly apparent to wheelchair users. Whereas for passengers who have their access needs met by the system the infrastructure can be said to fade into the background, wheelchair users need to tame the system into submission to their needs.

Infrastructure studies descriptions of infrastructure as invisible have rarely had to submit to these types of stories. Whereas we might be used to hearing about "broken" systems, we often hear of these thick descriptions when they are infrastructures of the Global South that suffer from lack of investment or maintenance.[5] We less often hear stories of systems that, one might argue, are entirely functional—after all, over one billion journeys occur in Transport for London's networks—but whose function is inequitably distributed due to its limited access-knowledge of users' needs. How, then, can we understand infrastructures through this complexity? It is unlikely that we can easily adjust the metaphor of invisibility to account for this, as it clearly fails to contend with misfit cases. For normate users, who can travel through the infrastructure in "relative anonymity," the infrastructure can feel seamless (Garland-Thomson 2011). But the phenomenology of the misfit is altogether different; their own identities are constantly being forced into visibility to the normates in the system as their relations with the infrastructure fall apart and as they put more and more work into making the system work for them.

INFRASTRUCTURAL INVISIBILITY AS READINESS-TO-HAND

The characteristics of infrastructures, as defined by Star and Ruhleder and largely used as the basis of analysis in various approaches to studying infrastructure, are closely intertwined and dependent on one another. Thus, to understand the proposition that "the normally invisible quality of working infrastructure becomes visible when it breaks" (Star and Ruhleder 1996, 113), we need to understand that the statement is predicated on two characteristics that come before it. The first characteristic is the embeddedness of infrastructure into other structures (social, technological, and otherwise), which is itself tightly linked with the idea that practices of an infrastructure are learned as part of membership. We have seen this through my development of the concept of designed materiality, which reminds us that an infrastructure's materiality is the result of entanglements and differential results of its social and cultural history that have shaped and informed not only the infrastructure's contours but also have defined its community of practice. Second, infrastructures are the embodiment of standards—through the use of regulations, design standards, and configurations that shape and afford different interactions and relations within them, infrastructures expect forms of behavior. Bringing these characteristics together, it becomes clear that if infrastructures are the embodiment of standards, they must therefore be the embodiment of standards that hold significance to a particular and specific membership (or, Star's preferred term, *community of practice*). Consequently, those who are not members of that community might find the infrastructure more difficult to grasp, requiring more work to connect the various threads needed to concretize the functionality for their needs. We have seen how wheelchair users were not members of the original community of practice of transportation (see chapter 1) and how they shape and mold the infrastructure through their use of belligerent techne forged by their access-knowledges (see chapters 2 and 3).

While the characteristic of invisibility is often offered in the introductory sections of articles and books on infrastructures, there has been limited explication of the specific intellectual histories that gave rise to this metaphor. Even less attention has been given to the analytical fallout of the invisibility metaphor.

This is particularly poignant given the caveat that Ruhleder and Star provided on the relationality of infrastructure. Steven Jackson has dutifully linked the characteristic of invisibility to longer histories grounded in both Deweyan philosophy of cognition and Heideggerian phenomenology of tool-being (Jackson 2014, 2015). In the latter, we have the distinctions between "ready-to-hand" and "unready-to-hand," the first of which refers to instruments used without having to make them work. I'll illustrate with the present case of transport infrastructure in London: the infrastructure is ready-to-hand when someone is at a bus stop, scrolling through their phone, then raises a hand as a bus arrives; it stops before them, they step onto the bus, beep their Oyster card on the reader, and find a seat or a spot to stand. They barely have to lift their head from their phone—the infrastructure is ready-to-hand because it requires little effort for each of the steps to occur. The subject-object relationship is virtually nonexistent in this context; it is only an experience of a task. When the tool required for a task is not available because it cannot be used (say, it is broken), then it shifts into unready-to-hand: one must think about the task and the tools needed for the task to accomplish it. Many scholars argue that there is no experience that is entirely ready-to-hand, but rather describe it as an idealized state (Cappuccio and Wheeler, 2010). I would tend to agree, given that even if there is a minimal lurch on the bus and we must grab onto the handrails, we are jolted into awareness that something isn't quite as smooth as it might ideally be.[6]

This allows us to return to Star and Ruhleder's point, which, I hold, is among the most significant and—in some regards—under-analyzed caveats of infrastructure studies. If infrastructures are fundamentally relational, all their characteristics are also relational—including invisibility. Invisibility, therefore, was never intended to be a descriptor of infrastructures in the general, functionalist sense. Rather, infrastructures can be understood to break down in different ways for different people, depending on their experiences, depending on their standpoints. If invisibility means "to break down," then invisibility itself becomes a question of one's position within the infrastructure. If you are a full member of the infrastructure's community of practice, a breakdown might be more obviously legible to all other communities of practice, but if you are at the margins, as Leigh Star often reminds us, the fault lines are more readily apparent. I will illustrate.

I argued in chapter 1 that the transportation system in London was designed to cater to commuters and pleasure seekers and was developed with arteries that connect train stations to one another as well as to key areas of London (commercial and financial centers, tourist areas, residential areas). The community of practice of users was largely understood as consisting of nondisabled persons heading out to accomplish some shopping, go to a museum, or reach to their place of work (or home from work). To that community of practice, the infrastructure is largely ready-to-hand: members can accomplish the series of tasks described above—put out a hand, board a bus, tap the Oyster card, settle for the journey (however crowded that journey might be).

Wheelchair passengers, however, are not members of this original community of practice. They have been newly included, with a budding but flattened disability access-knowledge now emerging in the system that believes that access icons are, in themselves, the creation of access. I showed the extent to which the infrastructure requires them to deploy their own knowledges and techne, mandating the weaving of useful strategies to smooth over the trip. For them, the infrastructure is fractured, fragmented, and incomplete without their own abilities to forge access. In this sense, the infrastructure must be *made* ready-to-hand for them through their own efforts. If the infrastructure is more difficult to apprehend to wheelchair passengers given their marginal position in the community of practice and, in this sense, the infrastructure's function of providing movement within the city is broken, we can therefore argue that it is not invisible. Quite the contrary: infrastructure is painfully visible to wheelchair users despite their desire for it to be invisible, precisely because it is so often and cumulatively fractured and dysfunctional for their needs. It requires significant practical problem-solving to surmount existing barriers and complex crip access-knowledges to circumvent the existing issues.

The invisibility of infrastructures is therefore necessarily a phenomenological concern: it resides in one's relation to and experience of the infrastructure. Indeed, nondisabled users of public transportation can likely easily recall moments wherein they were painfully aware of the infrastructure—the bus was significantly late, the train stopped on its tracks, they thought they had loaded money to the Oyster card but hadn't and got stuck at the barriers. These moments, however, are fewer and farther between than for wheelchair users.

One of the primary conceptual concerns that the invisibility narrative raises is, then, what are we speaking of when we speak of invisibility? Invisible to whom? Under what circumstances? I have argued that the metaphor of invisibility in infrastructures is relational. The infrastructure is "broken" for wheelchair passengers in London with much more frequency than it is for normate passengers. We discussed the infrastructure's historical imagined users in chapter 1. The purpose of the transportation network in London was not only to allow one to travel faster in a city that was already experiencing slow and heavy traffic. It aimed to tackle, specifically, the types of travel that the system builders recognized: work commuting and leisure activity in a time when disabled people were perceived to do neither. What is interesting is how this question creates a layered concern between the normative goals of the infrastructure in broader functional terms (i.e., public transport networks are public—for all—and for transport—to get around a city efficiently) and the social norms and mores of its time of conception (i.e., public transport was for a different public at a different point in history).[7]

We begin to see the clash between these functional and historical norms when our analysis centers on marginalized users' experiences rather than those of normate users. If we understand invisibility as readiness-to-hand, as I've argued is the common (if not always explicit) understanding of infrastructure, and we have shown how that readiness-to-hand is relative to the user, then we can argue that the process of going from ready- to unready-to-hand (i.e., breakdown) is the cessation of smooth operation. Yet, that breakdown is experienced not only according to the notion of function, but also to the appropriateness of the user to the infrastructure's design—whether they are an imagined user. Readiness-to-hand is, therefore, experienced further in relation to a socially endorsed norm. This explains why wheelchair users feel "invisible" to the infrastructure themselves! Indeed, for over a century, they were effectively invisible and unknowable to public transportation that understood them as neither members of the public nor requiring transportation.

There is, as a result, a gap between the functional norm that catalyzes the development of the infrastructure (public transport for all) and the social norm that is embedded, materialized, in the infrastructure through its original conception of the user ("all" does not include disabled people who were, through to the mid-twentieth century—and arguably are still—socially segregated and excluded). Infrastructures empower and enable human agents unevenly. I mean here that, for its imagined passengers, there is a clear fit between the function and social norm (i.e., "I want to get around London with ease; this infrastructure enables me to do so; this infrastructure works as it ought to"), whereas for non-normate populations (in our case, wheelchair passengers, though we can extend it to many diverse impairments) the function is barred through the social norm (i.e., "I want to get around London; this infrastructure does not understand *my* needs; this infrastructure is broken").

Therefore, the phenomenological experience of infrastructures is necessarily diverse, and this should be clearly articulated in our analyses of infrastructures when we say, uncritically, that infrastructures are invisible. Infrastructures are tools, meaning they are means to an end, and we see that for the imagined user group, there is a satisfying conclusion. Public transport in London, even if crowded, hot, and uncomfortable, is a means to the end goal of moving around the city. Heterogeneous user groups, marginalized groups, on the other hand, must be retrofitted in and even reverse engineer their own means into and through the network, making explicit the leading social norms that delimited (and, in many regards continue to govern) the functional norm. They make the gap between the normative functional and the materialized social visible, creating a disruptive reflection in infrastructural experience as well as a conceptual disruption in infrastructure studies.

Ultimately, this conceptual disruption, I argue, is a different kind of breakdown than that generally posited by those who refer to what has become a truism in infrastructure studies: that "any genuine infrastructure is mostly invisible" or, indeed, that infrastructures are invisible until they break down (Edwards et al. 2009, 370). It is, in some regards, also different from the oppositional stance that argues that infrastructures are mundanely visible because of the constant repair work that is required to keep them working. Rather, the conceptual gap that is highlighted through the experience of marginalized users demonstrates that breakdown is not just the cessation of an infrastructure's function in a general sense, but in a very specific normative sense as well. The infrastructure ceases to work if the users do not fit into the infrastructure's designed materiality, which, as we have seen, is itself the result of sociohistorical and normative assumptions and gradual accretion. In other words, visibility as unreadiness-to-hand is the result of a functional and normative crisis. To understand breakdown in this way enables us to highlight the value-laden historical circumstances that originated the development of the infrastructure and how that, too, influenced and constrained its definitions of functionality. It is the sociocultural and historical context brought in through a crip feminist lens that demands the additional concern of the infrastructure being invisible to whom and under what circumstances.

I have argued that breakdown cannot be solely understood as a technical cessation of operations that can be easily repaired and satisfactorily resolved through maintenance work. Rather, breakdown ought to be understood as the relational decoupling between function as an imperative (i.e., public transport is for movement across town) and to whom that functional imperative normatively applies (i.e., who is the "public" of public transport). This phenomenological and relational definition of breakdown allows us to denaturalize an infrastructure's function to not only query why it establishes particular functional norms but also ask through what means it attempts to attain them. This is always already questioned by wheelchair passengers, for example, in the ways that they express the injustice of transportation infrastructure in London.

> You just have to be super aware of where you're going, what you're doing, what time you need to be there, and how you're going to get there. And sometimes I think that's unfair because sometimes I just want to roll out of bed and go where I'm going, but unless you get in a cab, you can't do that.
> —*Alex Lyons*

As Alex Lyons argues, the fact that he needs to put in so much thought and effort into how he gets around on public transport is fundamentally unjust. He could not, even if he wished to, merely decide to embark on a journey via public transport because he knows it requires much more work for him to piece the network together, forge access, and make it work for him. Wheelchair users, in their crip access-knowledges, are profoundly aware of how fragmented the system is for them and their needs. Through their belligerent techne, they are laboring toward a functional system on their own terms. This normative gap between imagined users and marginalized users can only be tended to if we place experience at the center of our analysis and understand that infrastructures and all their supposed characteristics are ultimately always relational and dependent on one's positionality within the system.

In centering the experiences of wheelchair passengers who, I've argued, are marginalized users of this particular infrastructure, I have offered an alternative methodological inroad into studying infrastructure. It is a path that is different from Bowker and Star's (2000) "infrastructural inversion" and also diverges from Edwards's (2002) proposal for a multiscalar analysis. While the methodology here is, of course, central in that it has enabled me to highlight new perspectives on infrastructures, the deeper theoretical provocation that it has allowed is what I would like to focus on.

In proposing a multiscalar analysis of infrastructures, Edwards offers two important asides. In the first, he critiques social constructivist approaches for being necessarily micro-scale analyses and of a reductionist nature. In other words, the explanatory power always lies at the smallest level of the user. His point is to warn readers away from depending solely on the micro-scale analysis rather than to throw constructivism out entirely. I would add a further aside for macro-scale analyses. A macro-scale analysis done on its own develops a similar form of reductionism to the functional. As an example of a functional analysis, Edwards offers the process of supplementation and replacement of infrastructures whose purpose is communication, moving from postal services to the telegraph, the telephone, then email. The macro-level analysis suffers from a reductionism dependent on a naturalization of function that eschews broader ethical and social considerations. In the example of communication, it sees only the communications that are effectively made through the infrastructure being analyzed. In so doing, it is skewed to seeing normate users: those for whom those functions are made. It offers no sense of ethical considerations (e.g., Who can or cannot communicate?), privileging instead a myopic definition of function tout court. Ultimately, the macro view accepts function as the given, failing to tend to the social norms that materialize, enable, and constrain the identified function. The functional always happens against the background of the social, and an analysis that misses either, as Edwards correctly points out, will be reductionist in either direction.

Thus, a multiscalar analysis might begin to get at some of the concerns I mentioned, but one of Edwards's own imposed constraints gets in the way of its full realization. He offers, for example, that theorizations of infrastructure as "functional," here meaning "in working condition," might only truly be applied to countries of the developed Global North, whereas mundane breakdown is, too often, what defines infrastructures of the Global South. This is a simplistic, and unfortunate, generalization. A feminist technoscience framework further illuminates that distinction by bringing in the context of development of infrastructure in the Global South as the result of colonial legacies that have privileged specific population's needs over others, resulting in similar questions of normative assumptions embedded in these infrastructures' designed materialities. Furthermore, this holds just as much explanatory power in the Global North—my own case study being a clear example, but one need not go as far as London. Flint, Michigan, offers another example of how infrastructures in "developed" countries fulfill the functionality clause for some, but not for marginalized others.

However, the predominance of the invisibility narrative in critical infrastructure studies, as is aptly pointed out by Steven Jackson, might hide some of the "relations of value and order" that are hidden in our analyses of infrastructure (Jackson 2014, 231). However, while Jackson sees considerations for maintenance and repair as the way to correct this, I argue that it is in placing infrastructural experience, particularly of marginalized users, at the center that we can begin to peel away the layers of designed materiality. If, as I think is often the case with

STS work, we intend to develop theoretical tools that enable meaningful change toward a more ethical and inclusive world, our theories ought to denaturalize our objects of study all the way through. Our analyses of infrastructure need to keep infrastructures visible—not accepting function as smooth operation, for, as we have seen, that smooth operation is necessarily tinted with layered questions of social norms and values.

Ultimately, then, the idea of infrastructural breakdown and disrepair ought to be seen as experiential and relational. Doing so has consequences for how we understand and define the process of development of infrastructures over time and where to look for answers as to how we got to where we are. Importantly, it may also hold promise for how to think about reconstructive processes of infrastructures to ensure they fulfill their functional promises in more democratic and socially just ways, creating truly public infrastructures.

To attain this goal, we must learn from the diversity of experiences present in infrastructures. We know that a plurality of knowledges are being enacted and materialized within them (see chapter 3). Ultimately, the supported and legitimized inclusion of the knowledges of marginalized users, which they use to inform and shape their own interactions with infrastructures, may create real consequences for how planners and other users conceive of the materiality of infrastructures; how we repair, maintain, and reconsider our infrastructures over time; and potential paths forward for more ethical reconstructive processes of infrastructures in the future.

Conclusion / *"They're changing the network, the world, just by being there."*

In November 2021, nearly two years into the global COVID-19 pandemic and while I was in the process of writing this manuscript, I reached out to Alan. I told him that I was writing a book about the research he had participated in and that I would begin the book with a snapshot of our travels together. With his usual generosity, he congratulated me, and offered to conduct a follow-up interview. After all, so much time had gone by since we last spoke. It would be an opportunity to ask, What has changed since that trip?

This interview was quite different from our previous interactions, which had occurred in person, in London, five years prior.[1] We negotiated a time over a four-hour difference; I set up in my office in upstate New York and sent Alan a Zoom link—a process with which many of us have become all too familiar. The differences weren't only about distance, but how our lives had been profoundly changed in the past two years of sheltering in place and navigating each of our countries' varying responses to COVID-19. We managed, despite that, to focus our conversation on transportation and Alan's perceptions and experiences of how things had shifted in the five years since we traveled, particularly under the duress of the pandemic.

We began speaking broadly. Had there been improvements in the past years?

Yes, and no. Our theme of ambivalence prevailed. He explained how it is clear that significant money has been spent on accessibility issues in transportation. The number of accessible stations, for example, has increased since we spoke. Indeed, since I first conducted this research, the number of accessible stations has gone from 71 to 92 of the 270 stations on the Underground network, with a more modest increase from 57 to 60 of the 112 Overground stations. In 2022, Transport for London finally opened the Elizabeth line (formerly the Crossrail), which was originally to be inaugurated in 2018. Originally at least 7 of its stations were going to be inaccessible, but through the persistent activism of disability and accessibility advocates, all 41 stations will be accessible. This is, undoubtedly, a move toward a more accessible London. Alan also underlined how general attitudes toward disability and access in transportation has shifted in transportation companies with the establishment of access panels, or "groups of disabled people that they consult about policies, procedures, designs." He emphasized that another shift has been toward compensating the disabled people who are brought in for these consultation processes. Crip access-knowledge is gradually trickling into the infrastructure. There is some cause for optimism.

However, among his observations is that the quality of assistance for accessibility—such as staff training—has backslid significantly since the start of pandemic.

> Before all of that kicked off, training was beginning to get really, really very good. People were starting to see how important it is to involve disabled people. . . . But once the pandemic came along, two things happened. All of that face-to-face training disappeared, for obvious reasons. The other thing that happened is that most disabled people stopped traveling. Because staff then haven't been dealing with disabled people, they're not keeping that knowledge up-to-date.

What is of interest here is Alan's recognition that staff's knowledge and capacity for ensuring accessibility services is directly dependent on their interactions with disabled passengers on a quotidian basis.[2] It is through working with disabled people, through trainings or daily ridership, and becoming aware of disabled passengers' needs that staff at Transport for London and associated transportation companies acquire the forms of knowledge that are helpful and necessary for accessibility. It is as I have illustrated throughout this book: the presence, experience, knowledge, and visibility of disabled people is ultimately what makes this infrastructure otherwise. As Kerstin had put it to me when we spoke many years before this conversation with Alan:

> I think people should be aware that they're changing the network, the world, just by being there.

The changes have been gradual, but it has become evident (at least to disabled passengers who navigate the system) that when disabled people are actively included in design and decision making processes, the speed at which changes are carried out increases. So does the quality of the changes. That inclusion can be as seemingly small as disabled passengers simply—yet persistently—using the system. As Rosemarie Garland-Thomson reminds us, "The experience of misfitting can produce subjugated knowledges from which an oppositional consciousness and politicized identity might arise" (Garland-Thomson 2011). It is important not only that these subjugated knowledges be produced, but also that they are actively integrated into design processes at all levels for a truly democratic and socially just transportation system.

A key argument of this book is that infrastructure studies scholars need to push beyond the invisibility metaphor in their work. The crip feminist lens, in its commitment to ironies and to unearthing the knowledges of marginalized persons in infrastructures, enabled the crystallization of the question of invisibility as one that is necessarily relational. As such, it provides the added benefit of seeing how processes of stabilization and contestation of a system interact with and are the result of the engagement of marginalized users. Infrastructures are only ever "invisible" when the person interacting with the infrastructure is aligned with its normative imperative goals—including already being an insider of the system's community of practice.

The breakdowns and disruptions faced by wheelchair users are so constant that this infrastructure is *never* invisible to them. Their awareness of the variety of breakdowns that can occur and the collection of belligerent techne they have developed to intervene in these situations is evidence of this. It seems that there is no moment of passivity while using public transport as a wheelchair user, be it before leaving the house while having to prepare oneself for the trip or during the journey while remaining wary of one's surroundings, ready to deploy an array of ad hoc solutions. It is an interesting twist in the narrative that wheelchair users are such active passengers in transport, given the social stigma that often brands disabled people as passive members of society, unable to care for themselves (Finkelstein 1981). As such, wheelchair users in their engagement with public transport are often *at work*. My goal has been to surface this work, using the metaphor of niche construction as a way of highlighting it and recognizing the productive nature of wheelchair users' belligerent techne.[3]

It is through the niche construction of the infrastructure that wheelchair users are making themselves visible to the system. Alan had already highlighted this to me in our original interview:

> By being out and visible, I'm actually going out, like lots of other disabled people are, and changing people's opinions and impressions.

In other words, while traveling, wheelchair users are physically visible in the system. This, in and of itself, plays an important role in demonstrating two things. First is that wheelchair users, and disabled people in general, *do* use public transport, and that, therefore, more accessible services need to be procured.

Alice, another interviewee, described an interesting negative feedback loop in which few accessibility provisions are given, allowing low numbers of disabled people to use the service, thereby causing service providers to justify a lack of investment in additional accessible services through lack of use:

> I know about that kind of thing that goes on in the minds of transport providers where they go, "Ah, but not many people, dot dot dot. So we're not going to make much provision." And then that very provision drives the behavior of wheelchair users and when they first brought in the low-floor buses with ramps in London, there were about two buses and there were people saying, "But nobody's using them." And I said to them, but you can't go anywhere, you might be able to get on a bus somewhere but you can't get off it again.

Alice, who has experience working and consulting for transport providers, knows what they think: it does not seem like many disabled passengers use the provisions made for them. However, she disagrees with the causal relationship. The reason why few wheelchair users are using public transport is not because there are not many of them, but because the provisions are insufficient and limiting. Interviewees identified that, in order to break this cycle, one must go out and use the public transport infrastructure despite its inadequacies and be seen doing so. *Becoming visible.*

By being present in the infrastructure, using it, wheelchair users make not only themselves visible, but also the infrastructure itself. These are moments in which the cloak of routine is stripped away to other passengers using the infrastructure, revealing to them the gaps that marginalized users encounter and bringing the bare bones of the network to the foreground. As a result, we come full circle in the tensions of visibility: two entities, socially invisible to normate users, become visible through their interaction. On the one hand, as disabled people, wheelchair users have for centuries been stigmatized and marginalized, othered and segregated, cared for in private spheres. As a result, they became socially invisible and absent. On the other hand, through processes of standardization and stabilization, transport infrastructure has become invisible to its core community of practice, only revealing itself in moments of breakdown. In interacting with the system, wheelchair users break the silence and insert themselves into the narrative, highlighting the inadequacies of infrastructure, the limitations of its presumed public, thereby creating a normative crisis. It is perhaps with this in mind that, when asked what words of advice they would give to a wheelchair user attempting to travel using public transport in London, so many interviewees prefaced other advice with a simple "go for it" (Diana).

Infrastructures are tense creatures. They have the power to exclude users but are still malleable enough to allow subversive points of entry for change. The notion that infrastructures are invisible is a privileged perspective—indeed, it is a definition of infrastructure that can only be made by those who are fully, or at least mostly, inscribed in a given community of practice. To say that public transport in London is an "invisible infrastructure" is to throw by the wayside the experiences of wheelchair-using passengers. It is a claim that can only be made by those who are fully, or mostly, inscribed in an ableist societal structure. For those at the margins, whatever those margins might be, there is a constant awareness of the various ways in which infrastructures demand additional work on their part in order to make infrastructural use possible. And many people are willing to recognize this when they are themselves placed at the margins. Conversations about my research with friends almost invariably surfaced stories of fraught experiences with public transportation when they had, for example, a broken leg due to an accident, and how much harder traveling became.

Without the crip feminist lens, this would have been difficult to establish. Indeed, the narrative may have become one of linear progression, changes affected by technological advances and adjustments to higher demands or external factors such as wars (in the same way that the First and Second World Wars closed Tube lines) or petrol prices. By concentrating on the experiences of excluded users, infrastructures here have acquired depth and complexity, showing how their edges are defined by specific considerations of who their users are and consequentially placing barriers in the path of those they do not see as part of the system. Yet the story does not end there. This is not a story of being left at the borders but one of entrance; not a story of passivity but of activity and activism, exclusion and inclusion. It is through choosing to listen to the stories and experiences of wheelchair users that we see that infrastructures can be pushed, prodded, molded by those to which it had not previously catered. This, then, is a story of inversion—not infrastructural, but social. In giving space to the voices of wheelchair users, we have turned the usual social narrative of them as passive members of society on its head, recognizing their unique lived experiences and their impact on infrastructures, shaping the latter in innovative ways.

NICHE-CONSTRUCTING DESIGNED MATERIALITY

Developing the concept of designed materiality enables me to crystallize a focus in the sociocultural histories of infrastructures to show how they are the layered result of shifting social norms that have consolidated and congealed over time. Ultimately, this concept also helps locate the work of marginalized users in the infrastructural ecology as it can slowly shape and mold the built environment. In focusing on the form of materials—the choices made in how the assemblage is put together, its affordances—I illustrated why it is that the disability access-knowledge of the infrastructure is limited. I critiqued acts of retrofitting as not being the restorative, reformative, bridge-building acts that others have argued that they are. Rather, retrofits are an extension of the infrastructure's legitimized knowledge practices and often extend the infrastructure's limitations in new and unpredicted ways.

I contrasted the act of retrofitting, as the infrastructural response to new demands, with the work of wheelchair passengers, and I argued that the latter might be understood through the metaphor of niche construction. If an infrastructure can be argued to be a niche, with various groups of actors working in it, the work and knowledge of wheelchair passengers in it has material consequences as well. We have seen the niche construction of wheelchair users in various ways, both as transient, specifically generational acts through the activism of the 1980s and 1990s and through the persistent, intergenerational belligerent techne these users have developed and shared. Ultimately, in picking up Leigh Star's argument, our world could not just be otherwise: it is often *made* otherwise by marginalized users.

The process of including marginalized users is therefore currently dependent on these users mobilizing their own forms of access-knowledge. I illustrated the layered and complex forms of knowledge that wheelchair passengers have of the transportation infrastructure through a blended narrative. It captured the decisions, cognitive processes, and deep experiential practices of wheelchair passengers as they navigate the system from choosing where to live, which wheelchair to use and when, the tools they carry to hack the system into a functional state. Wheelchair passengers' various forms of know-how are often working in direct opposition to how the system wants them to behave. This push and pull of charge and control was mapped back to the infrastructure's regime of disability access-knowledge—where disabled passengers' needs are reduced to categories that are often poorly matched to their actual needs, and priority debates that extend social hierarchies—that is contrasted with wheelchair users' own crip access-knowledge. Crip technoscience brings to the foreground how, in their daily interactions with the system, wheelchair users begin the process of reshaping the infrastructure and taking back control. However, if we are to change the stasis that Alan has described in the past few years in public transport in London, we must forge new paths for legitimizing these passengers' knowledges. As the situation currently stands, it is clear that their knowledges are still marginal and that they are still dependent on their belligerent techne to forge access. New paths forward for including diverse knowledges in infrastructural development are required for creating more socially just networks.

FROM NICHE CONSTRUCTION TO TARGETED UNIVERSALISM

Ultimately, the question that remains at the end of this book is of what can be done to accelerate the pace of disabled persons' niche construction of the transportation infrastructure. Or, perhaps more precisely, what can be done to ensure that the actions of wheelchair users do not have to be repeated in such ad hoc manners in order that their belligerent techne become integrated as permanent parts of the infrastructural ecology of transportation in London. In chapter 2, I argued that niche construction is a promising metaphor, for it not only describes the ways that wheelchair users shape the infrastructural environment but can also serve as a way of identifying feedforward loops for more socially just systems. If crip access-knowledges are accepted as a form of knowledge that ought to be incorporated into the system, as I advocate, it is through the various tactics that wheelchair users deploy that designers, engineers, and policy makers will be better able to identify paths forward for accessibility. In this, through the crip feminist lens, this work further serves the purpose of collecting data on the experiences of marginalized users and can function as useful evidence to present to institutions that may have stronger influence in speeding up the process of changing infrastructures. Bringing designed materiality to the heart of the matter, and showing how it enables the analysis to highlight user experience, we must demand new design practices that see that, when a design is being made, one needs degrees of freedom, flexibility, and belligerence. We need the space for possibilities.

The concept of targeted universalism was developed by john a. powell, director of the Othering and Belonging Institute at UC Berkeley.[4] powell critiques what he calls a *false* universalism in the face of racial inequalities in the United States, particularly the notion that neutrality of design creates neutrality of effects (powell 2009). He offers the example of the unequal effects of the GI Bill whereby Black veterans, while technically receiving federal money to attend college upon returning from war, attended colleges that were highly segregated and of poorer quality. Extending notions that contrast fairness as equal treatment for all with equity as improving the conditions of those in more disadvantaged positions, powell argues that the only way to redress social inequalities in public policy is through targeted universalism. This, he articulates, is a concept that is "inclusive of the needs of both the dominant and the marginal groups, but pays particular attention to the situation of the marginal group" (powell 2009, 803). It rejects a one-size-fits-all approach to policies, while embracing the differences between affected groups. Thus, for example, if a universal aim is to eradicate poverty, targeted universalism would demand sensitivity to the communities that have been intersectionally, culturally, and socially marginalized, such as Black and Indigenous peoples of color, LGBTQ+ communities, and disabled populations, depending on their positionality with respect to the aim. More precisely, targeted universalism is an outcome- and goal-focused approach to policy making and decision making processes. It aims to establish universal goals and advance the entire population toward those goals. However, in recognizing that not all persons are equally situated in society, it asserts that in order to achieve those goals we must advocate for multiple paths toward them—in other words, create "a range of implementation strategies" (powell, Menendian, and Ake 2019, 15).

If infrastructures have normative goals, even implicitly, then they lend themselves particularly well to the targeted universalist approach in both design and policy making. Public transportation, with the goal of providing mobility to the public around the city, would have mobility for all, including disabled persons, as its universal goal.[5] We understand, as has been argued throughout this book, that not all persons are on equal footing in relation to the infrastructure. Through its history, transportation infrastructure has established unequal access that requires redress. Having a diversity of strategies that advance the interests of marginalized populations toward the universal imperative would enable us to reach that goal more speedily. The identification of these diverse strategies might begin through work such as this that identifies the knowledges that marginalized populations already have of their own situations. In recognizing the knowledge and work of wheelchair passengers in transportation, one may offer strategies that either work toward redressing them having to do that work or that incorporate those strategies permanently into the system.

Targeted universalism is, in this sense, directly allied with disability justice and may offer disability activists paths forward for articulating their needs and demands in ways that become legible to institutions. In the context of infrastructure studies, a targeted universalist approach would enable a more nuanced discussion of infrastructural aims and its functions while still grasping the complexities put forward by designed materiality and pushing beyond the invisibility metaphor—that is, that infrastructures distribute their function inequitably, but we can understand their functional imperative to still be universally desirable. Targeted universalism in policy planning for infrastructures can therefore accept that there are sociohistorical contingencies that affect heterogeneous populations in different ways while still describing an infrastructure's goal (be it transportation, access to water, or access to the National Health Service) as universal in nature. If the designing mind moves away from a notion of infrastructure as a stable, functional, materially enduring object and instead adopts a perspective that infrastructures nest and fit into existing social niches, it may enable the bringing in of new knowers, new epistemologies, and new techne. Bringing in the belligerent techne of wheelchair users would require, for example, a permanent shift in the infrastructure's designed materiality.

Ultimately, the only manner through which we will create more socially just and democratic infrastructures and worlds is through the legitimization and incorporation of diverse knowledges. Through arguing that the work that wheelchair passengers do in transportation is a form of molding that infrastructure, I want to emphasize how this work is already ongoing in subversive and unexpected ways. If we are to enable these processes to occur not only as the truant acts of freedom of marginalized populations, then we must also establish ways to respect, legitimize, and recognize crip technoscience as doing the work of depetrifying our futures.

As my 2022 conversation with Alan began to wrap up, I asked him one final question: What do you see as being the future of transportation in London? His answer to that question is the perfect place to conclude this exploration. It often was the case in my interviews that wheelchair users' articulations elegantly captured the essence of what I try to convey in studying infrastructures. It doesn't matter what type of infrastructure we are ultimately analyzing, whether transportation or energy systems or waterways. The experiences that diverse populations have of these infrastructures are our entryway to understanding how to improve our worlds. In closing, here is Alan:

> I've got to assume that the future is equality of experience. If I thought that it wasn't possible to get . . . to give disabled people . . . I only want the same miserable experience as every other commuter. If I didn't think that was possible, I don't think I would keep doing what I do.

Notes

Introduction

1. Quotations from interviews will be attributed in one of two ways: in parentheses after the quote or as block quotes in separate paragraphs. The work here is the direct result of interlocutors dedicating time and energy to speaking with me, inviting me to spend time with them in different contexts, and their voices ought to be clearly attributed.

2. Data is from January 5, 2019, to January 4, 2020, before the COVID-19 worldwide pandemic truly hit the United Kingdom. The London transportation system was significantly impacted by the pandemic, with passenger numbers plummeting to 1.8 billion journeys during 2020. At the time of writing this note in the first draft of the book (December 2021), this number seems to be slowly recovering, as the United Kingdom has been hit by multiple surges of COVID cases throughout 2021. Some reflections on the impact of the pandemic on transportation accessibility are offered in the conclusion of this book.

3. I see the subfield now often referred to as "infrastructure studies" as having emerged, at least in part, from Thomas Hughes's historical work on sociotechnical systems. I seem to be supported in that assumption by Paul Edwards. His chapter "Infrastructure and Modernity: Force, Time, and Social Organization in the History of Sociotechnical Systems" points to the interchangeability of the terms *sociotechnical system* and *infrastructure* (Edwards 2002). Thus, and so as not to repeat the term *infrastructure* ad nauseam throughout the text, I use the terms *infrastructure*, *system*, and, more rarely, *network* interchangeably.

4. *Buggy* is a British term for a baby stroller; another term for it commonly used in interviews is *pushchair*. I will be using these various terms interchangeably, betraying that my English is from neither here nor there, as I acquired it a little bit everywhere.

5. This has meant, in particular, taking the time to create summaries of my research and to remain in communication with a large number of interviewees who welcomed that interaction. The former occurred for both my master's and doctoral work, each dissertation being summarized in a pamphlet format (not exceeding fifteen pages) distributed via email to participants and stakeholders (including Transport for London). Both documents are freely available on my personal website. The latter has largely

occurred via Twitter, where interlocutors and I have remained in touch as mutual followers and via direct messages.

6. At least two individuals present as gender nonbinary and use the pronouns they/them/theirs. This was confirmed via more recent social media interactions. I was unable to reach all interviewees to ask whether their gender expression has shifted in the years since we sat down for our original conversation and have maintained the pronouns they asked me to use at our first meeting.

7. Some interviewees did choose to disclose this information and discuss their abilities as directly linked to a particular diagnosis. Where freely offered in a relevant quote I have included such disclosures, as participants had the opportunity to read through interview transcripts and redact any information they did not want used in research.

8. I have chosen not to make a clear distinction between pseudonyms and real names. Some cases may be clarified and others not.

9. A disability roadshow, titled thusly by Marie, an interviewee, is an event where disability organizations are invited to spend time at a transport provider to speak directly with employees such as drivers. Marie, a wheelchair user, organized one at the bus company she works for, with the aim to show employees the "human experience" of access, which she deemed preferable to receiving email reminders about company policies regarding accessibility.

ONE / Partial Histories

1. I have, as I am sure readers will appreciate, significantly summarized centuries of history for the sake of conciseness while still hitting the key notes relating to the pattern of development of infrastructures as I identified them through various tomes on the history of public transportation in London.

2. It is unclear where users, for example, might figure in Hughes's approach as they do not seem to be either system builders on the inside or part of the environment on the outside.

3. Like McRuer, I want to resist saying that disabled people were relegated to the domestic realm because that space, too, as he notes, was one of compulsory able-bodiedness and heterosexuality. Disabled people were not included in the domestic realm but rather medically treated and cared for in that realm, not perceived as active participants in their own lives. Further, I have opted to include *housewife* as a category consistent with those that the transport system was being designed for, as women were targeted users of the system, seemingly to partake in the newly established department stores of central London. For example, Martin describes posters depicting "ladies in white dresses" using the Central Line, which opened in 1900. Martin mentions that this line ran extra evening trains for shoppers in the 1960s (Martin 2012).

4. One example illustrates this movement rather well. The City of London, the city's financial heart to this day, experienced a population decrease between the 1830s and 1850s. Today, it is largely a ghost town outside of business hours. Meanwhile, St. Giles, just east of the city, was reported in 1836 as having "260 houses with an average of 20 people in each" (Barker and Robbins 1963, xxvii).

5. Were this book a detailed history of public transportation in London through a large technical systems approach, I would trace the invention of the omnibus to Nantes, in France, by Stanislas Baudry (Vance 1986) and account for the shifts in the process of technological transfer through what Hughes would call *technological style* in its further adoption in Paris and then London by George Shilibeer. But this is not within the scope of my proposal here, which is to sketch out the broader pattern of infrastructural development over time and then problematize this common narrative.

6. The cut-and-cover method is aptly named. Workers cut a road open, dug a trench into it, and covered it up again, creating the underground tunnel. This method had its limits, namely the need to either cut open roads (thereby affecting traffic) or buy the properties through which a line was meant to go. These shortcomings would soon be solved.

7. To simplify (and clarify) the story, I use names of the modern lines rather than their rail company predecessors'.

8. Americans might be familiar with Yerkes because he was involved with the Chicago transportation system (such as the famous Loop), including more than one instance of blackmail. He is credited by University of Illinois archivist and his biographer, John Franch, as having "perfected" corruption in Chicago (Franch 2006).

9. Some famous iconic stations still in this style are Covent Garden, Chalk Farm, and Russell Square.

10. In his history of London transport, Ben Pedroche gives 1911 as the date for the first escalator in a London Underground station, and asserts that its success led to the gradual replacement of elevators in many other stations (Pedroche, 2013). Some could not take the replacement, as escalators require more space than elevators, hence why some stations (such as Covent Garden) are still dependent on elevators or stairs. He further notes that the process of replacement actually required significant remodeling of some stations, some of which resisted.

11. Only some streetcars now survive in London, where they are now known as the Tramlink system in Croydon (south London). They are not simple historical remnants but have their own complicated history of being closed and then redeveloped and reopened in the 2000s. As this system is specific to that area, it is not of significant focus to our broader story, especially as its operations opened in the twenty-first century.

12. Perhaps, if we must point to any single technological innovation or development of this era, we would choose the motor buses. Vehicles became larger, and covered

tops were allowed starting in the mid-1920s. Their seating capacity trebled and their speeds increased. Buses in London were already predominantly painted vermillion and the standardization process had already begun under the London General Omnibus Company, which built the X and B types in its own garages. In 1912, the manufacturing part of the London General Omnibus Company became the Associated Equipment Company (more broadly known as AEC), which eventually designed and built the famous Routemaster in the 1950s (though, to a nonenthusiast of transportation history, they might not look all that different from previous models such as the Regent III RT).

13. Brendan Gleeson argues that the lack of work in disability historiography was a serious gap in the field of disability studies. There has, of course, been some significant work developed in that vein in the past decades (Gleeson 1997). A couple of key contributions are *The Oxford Handbook of Disability History* (Rembis, Kudlick, and Nielsen 2018) and the New York University Press series on the history of disability, with a focus on the United States, edited by Paul K. Longmore and Lauri Umanski. However, few historians have attempted to directly discuss historical overlaps between the development of the built environment and disability history. The work of Bess Williamson and Aimi Hamraie stand out in this vein, but both have focused on the United States and specific considerations regarding design (Williamson 2019; Hamraie 2017). Many of their insights are directly applicable to my own work in later chapters (see chapters 2 and 3).

14. The Poor Laws, while related to, are not explicitly the same as what, in the United States context, is known as the Ugly Laws. As historian Susan Schweik traces in her book, the Ugly Laws have "cultural seeds" in the British context, including the persecution of the "deserving poor" if they are deemed to fall under "unsightly beggar ordinances." However, as Schweik further elaborates, the Ugly Laws have a "peculiarly American grain" in the ways they were phased in and enforced by municipal ordinances (Schweik 2009). I thusly emphasize these less than what seems, to British historians, to be the main thrust in disability history: charity, institutionalization, and segregation.

15. This depiction has recently been problematized by Daniel Blackie, who argues these claims are empirically weak; he is aligned with some of Gleeson's critiques on lack of historical work on disability (Gleeson 1997). Blackie's work on the coal industry in the Industrial Revolution in Britain is a particular attempt to redress some of the gaps in disability historiography, especially in relation to the presence of disability in the industrial workforce (Blackie 2018), but the gap largely perdures.

16. Colin Barnes and Mike Oliver have problematized the current state of disabled people's movements. They recognize that there have been significant legislative advances but argue that these, too, may be limited. In the 1990s, these authors were already cautious of an overly close relationship with the government, fearing that the

movement might be appropriated and manipulated by political interests. In 2006, prior to the absorption of the Disability Discrimination Act of 1995 by the Equality Act in 2010, they argued their fears were justified. They saw malice in the creation of the Disability Rights Commission as a way for the government to enroll disability organizations that would defend the status quo rather than fight for the rights of disabled people. They affirmed, "We no longer have a strong and powerful disabled people's movement and the struggle to improve disabled people's life chances has taken a step backwards" (Oliver and Barnes 2006). Barnes and Oliver argued that improvements in the daily lives of disabled people have been "more apparent than real" over the past decade, a statement corroborated by the think tank Demos. In 2006, Demos published the report "Disablist Britain," concluding that disabled people still experienced discrimination in contemporary British society (Miller, Gillinson, and Huber 2006) and that, worryingly, disability discrimination is an under-researched topic. It is now fifteen years since the Demos report and disability discrimination is still an often unspoken issue, left out of many debates around identity, inclusion, and equality.

17. While the concept of splintering networks (Graham and Marvin 2002; Guy 1997; Kooy and Bakker 2008) helped account for the process of privatization, it is clear that the specificities of particular systems cannot always be accounted for in general patterns of evolution. Indeed, Transport for London still is splintered in many ways, with various contracts being made with private companies to run bus lines as well as manage staff of the Underground and Overground lines and now the Elizabeth line. The organizational hierarchical mapping of Transport for London has changed multiple times in the past twenty years. While it has definitely "splintered" in some ways, it has also, in fascinating ways, coalesced into a single entity.

18. The most recent version of "Your Accessible Transport Network" is titled "Accessible Travel in London" and is introduced by Alice Maynard, board member of Transport for London and wheelchair user.

19. This included the creation of a new type of manual ramp to be deployed in cases where the train is lower than the platform. Train floors are usually higher than platforms in London stations.

20. S-stock trains were a welcome change for various reasons: they have four designated wheelchair spaces aboard (contrary to the usual two) and they were the first air-conditioned trains introduced in London. "Neapolitan" is an affectionate nickname for the grouping of the Metropolitan, Circle, and Hammersmith & City lines, represented respectively by magenta, yellow, and pink lines on the London Underground maps. These lines service the same stations along a central artery of London, from Baker Street in the west to Liverpool Street in the east, a stretch that is represented by the three colors combined and looking suspiciously like Neapolitan ice cream (chocolate, vanilla, strawberry).

21. The latest time of revision of this number was July 2023. The number of step-free stations on TfL went up by a total of eleven during the process of writing this book—a somewhat optimistic way of keeping track of the time it took to write a book!

22. This term will be familiar to UK and US audiences as either "reasonable adjustment" or "reasonable accommodation." Alex Lyons, an interviewee for this research who is a young white man in his twenties who has studied law, referred to the reasonability clause as the "right to discriminate"—a way of interpreting it that has stuck with me over the years.

TWO / Designed Materiality

1. I was fortunate to be Langdon Winner's colleague at Rensselaer Polytechnic Institute for four years prior to his retirement. It seems worthwhile to preserve in this note that, though it is perhaps not widely known, Langdon Winner was an instructor and supporter of RPI's design, innovation, and society major, which celebrates twenty-five years of existence in 2023. He taught the seminar class "Design, Culture, and Society" for many years, and I firmly believe that this department has been ahead of the curve in STS thinking about and with design.

2. This is how Winner's argument is developed. Of course, one must also point to the work of other scholars who have problematized Winner's account of the "racist bridges," including the very premise of *intention* (Woolgar and Cooper, 1999).

3. This government official spoke to me on condition of full anonymity, and he requested to be referred to by this vague moniker. He is a nondisabled man who has worked directly on accessible transportation in the United Kingdom.

4. Some newer buses also include a visual announcement on the dot matrix board (on the ground floor, positioned over the wheelchair priority area, and on the second floor of double-decker buses) that reads, "Wheelchair space requested," and an automated audio announcement that says, "A customer needs the wheelchair priority area. Please make space." When I undertook these interviews, however, these buses were few and far between, so the wheelchair users quoted here are all referring to the loud siren noise and not other indicators. This surfaces some interesting questions about how and when this change was made, but that is unfortunately outside the remit of this book due to the timeline at hand.

5. I am grateful to Aimi Hamraie's work and their book *Building Access*, where, in chapter 7, they discuss the debate around retrofitting and universal design and cite Jay Dolmage's scholarship.

6. Lauren Berlant passed away as I was writing the first draft of this chapter, and I am grateful to have the opportunity here to write a note of gratitude for their important scholarship.

7. This campaign has been "renewed" in a couple of ways. At a disability roadshow event I attended, a Transport for London bus representative told me about the "shelf life" of campaigns and that they were working on renewals. The first renewal came in 2015, when the colors of the original poster were inverted (it was originally red with white letters, but there were refreshed versions that were white with red letters in 2015). The second renewal came as part of a Mr. Men and Little Miss "behavioural campaign" on public transport (Loveday 2016).

8. Exceptions in STS that include some evolutionary metaphors include the work of some historians of technology and some Dutch schools of science and technology studies. For a useful literature review, see Schot and Geels (2007).

9. In the biological parlance, scholars refer to a modification of the selection pressures. However, as a metaphor, I think we can be satisfied in understanding it as the modification of the infrastructure in ways that are experienced not only by the user group that enacted the change but by other users as well.

10. I am so grateful for the work of Stacey Milbern, who is now herself a disabled ancestor. May she rest in power and may we all continue to learn from her wisdom.

11. I discuss here solely the requirements for rear-facing wheelchairs on public service vehicles as this is the paradigmatic arrangement on London buses. There is some (minimal) literature on the guidelines surrounding forward- versus rear-facing options for wheelchair spaces, but due to space and time constraints, I will keep this to the footnotes.

THREE / Situated Knowledges

1. At interviews, I always wrapped up our conversations with three "rapid-fire" questions as a manner of signaling that the interview was ending and introducing what I hoped would be a more lighthearted tone after at least an hour of rather heavy topics. The questions were, "How would you describe wheelchair access in London in three words?" "What are two areas of access that need the most work?" and "What words of advice would you have for a wheelchair user starting to navigate London public transport?"

> Among the many wonderful responses to these questions, perhaps my favorite was when Marie Claire asked if they could give me four words describing wheelchair access and said, "Not very accessible yet." It stuck out to me as an incredibly realistic and also *hopeful* response with the sly inclusion of "yet."

2. A thing of a bygone era, it seems! It would be curious to think about the process of retrofitting these large, stubborn keys that want to remain at the lobby with the newer key cards that nestle into wallets for months after a trip.

3. It is also a good space to remember that being a wheelchair user does not necessarily mean that the person does not have the ability to walk at all. There are many ambulatory wheelchair users who might use wheelchairs as assistive devices for a wide range of reasons.

4. Taxis in London are overseen by Transport for London, like all other modes of transport discussed here. Perhaps due to the door-to-door, individual nature of taxis, informants often spoke of them as nearly separate entities.

> Another transport alternative that was mentioned by three interviewees was cycling. Basil and Char Aznable mentioned it briefly, and Sophie is associated with a charity that campaigns for more widespread and accessible cycling options for disabled people. This is an intriguing new possibility in terms of transport, but it will not be further expanded in this book.

5. A few years after the research for this manuscript took place, I was sitting in my office at Rensselaer and scrolling through Twitter when I came across a @TfLAccess tweet saying that TfL was consolidating its Twitter accounts—they even had separate accounts for each Underground line—to a singular one. The reaction was swift: disabled people quickly created a Twitterstorm on this, pushing back against TfL's decision, speaking to how much they use the dedicated access account. Disabled people's swift reaction succeeded in getting results. Ultimately, @TfLAccess is one of three accounts that survived TfL's Twitter accounts purge thanks to collective action.

6. It is also interesting to note that, out of the seven interviewees who spoke of these skills, five identify as men. This could be due to the sample of interviewees—73 percent of men interviewed (eight of eleven) are manual wheelchair users as compared to 43 percent of women (six of fourteen). Chiara identifies as agender and uses a manual wheelchair.

7. We will discuss these victories again in chapter 4 as a part of a bigger feedback loop that currently exists in infrastructures and discussing how "access victories" can become disability technoscience itself, as it clearly has in many of these cases. I leave here a brief note to point to how this is, in no small part, due to infrastructural language being that of standards and categories.

8. Some notable successes for Transport for All include saving the boarding ramp scheme across the network and ensuring that all Crossrail (now the Elizabeth line) stations would be step-free when the line opened in 2018 (originally, seven stations were to remain inaccessible). I should also note that the Elizabeth line actually began operations in 2022, four years behind schedule.

9. Complaint logging used to be a significant part of what TfA did in the mid-2010s. Their website even had a sidebar titled "Your transport complaints" in which they provided a complaint form and guidance on "how to complain effectively." They still

provide advice on this, but seem to have shifted away from this explicit role toward more general advising services.

10. This campaign seems to have been a particular effort in the years 2014–2018.

11. We may borrow, from the design world, the terms *affordance* and *interface*. Both of these ideas get more to the ways in which designers think about design work in relation to the user's experience. Ingram et al. (2007) argue that these terms are more shallow than *scripting*, which, they say, is "more subtle"—with scripting, humans are treated "as social agents capable of resisting" (9). Their point is taken, but I believe this to be a somewhat ungenerous reading of design traditions and disciplines that have also come to reflect on the relations between users and artifacts through affordances in more critical ways.

12. The map may be found on the Transport for London website at https://tfl.gov.uk/transport-accessibility/download-accessibility-guides-and-maps. I heartily recommend that readers look at it while reading this section. Note the caveats associated with the accessibility features: the gaps sizes and number of steps between platform and station are measured at "these points only," and "lifts may not be managed by Transport for London."

13. Out of personal nostalgia, I often pick stations for hypotheticals that hold meaning to me. Funny how, despite the many years away from London at this point, the London transit system has a hold on me like no other object of study ever has. I offer here the journey I took many times from my home to my alma mater, University College London.

14. See chapter 2 for more discussion of the level of detail of these regulations.

15. The common Tube map signals wheelchair accessibility in broad terms compared to the detailed step-free Tube guide: it uses a white wheelchair user icon on a blue background to symbolize step-free access from street to train and a blue wheelchair user icon on a white background to symbolize step-free access from street to platform. The "white" symbol Anton is referring to is the latter, meaning access there requires a manual boarding ramp.

16. It should be noted that the Transport for London maps have since been updated to reflect this possibility. When we did this trip in 2015, there were no manual boarding ramps available at Hammersmith station for the District and Piccadilly lines. They have since been introduced, and this interchange between lines has become legitimized.

17. This is true as of 2021! When I first embarked on transportation research in London, I had to write a caveat, but no more. Transport for London used to run two "Heritage" lines (operated by Stagecoach London), routes 15H and 9H. These still operated with traditional "hop-on, hop-off" Routemasters (arguably the most iconic of London's buses, with the permanently open back door that allowed passengers to

board and alight whenever the bus was stationary) that had steps. The 9H line was permanently retired in 2014, and the 15H, after a brief stint as a "seasonal" Heritage line, was finally cut due to the COVID-19 pandemic.

18. There isn't any debate around this topic. A fascinating "Synthesis of Transit Practice" titled "Use of Rear-Facing Position for Common Wheelchairs on Transit Buses" was written by the US Transit Cooperative Research Program in 2003. The synthesis offered some conclusions on the use of rear-facing seats across Europe and Canada, including benefits for wheelchair passengers (ranging from "independent and dignified use of the system" to "less damage to the mobility aid") and to the transit system. I do not debate the findings of the synthesis here, but want to point to the ways that these syntheses are still largely written and produced by "practitioners" of transport, where the views of wheelchair passengers are but briefly taken into consideration. For example, the report points to "some users [feeling] singled out by being the only passengers facing to the rear," but suggests that this can be easily minimized if other passengers also face the rear. It was not the report's goal to investigate further than that, and I suspect that they are at least partly correct. That said, the question of *choice* is also an important one. If one can *only* face the rear, regardless of whether others are facing the rear, too, there isn't much of a choice offered.

19. The minute details of the case are not particularly relevant here. Nevertheless, readers may be interested to learn that the Supreme Court's ruling was rather ambiguous, though largely in favor of Paulley. In sum, the court ruled that the bus driver had not gone to reasonable lengths as they did not ask the parent with the buggy to vacate the area. Furthermore, the court ruled that drivers should bear more responsibility in requesting that the wheelchair area be vacated, which is, in itself, an interesting question for labor issues. Is it fair, for example, to place the burden on the driver who already has many other pressures to meet (e.g., keeping to schedule)? Who else can be held responsible?

20. Indeed, the Big Red Book of 2014 clearly stipulated, "Wheelchair users are to be given access to the wheelchair priority area even if it is occupied by other passengers or buggies. Use the iBus automated announcement to make it clear that the wheelchair priority area is needed." That said, it does not provide much further guidance to the driver should they come across someone who downright refuses to make space. As mentioned in note 19 above, this is a murky legal area.

21. The "new Routemaster" is a new model of bus that was heralded by former mayor Boris Johnson. Perhaps not-so-lovingly nicknamed "the Boris Bus," it was said to be, among other things, a Johnson vanity project.

22. It ought not to be left unnoted, as I continue to describe these tactics, how one might also analyze the tactics, their success, and the wheelchair passenger's other intersecting identities. For example, a tactic that might work well for a straight white dis-

abled British man in a manual wheelchair might completely backfire for an immigrant queer disabled woman of color. These intersections are important, as not all persons are interpreted or socially received equally in the contexts that they inhabit.

23. Honestly, the priority point ought to be moot. The infrastructure itself names it the "wheelchair priority area," but were I offering here a more detailed analysis based on crip theory, I might further elaborate on the question of compulsory able-bodiedness and compulsory heteronormativity, showing the parallels in both and why, in social contexts, a sexually reproductive person with a child can be seen, and see themselves, as more important than a disabled person. This is out of the remit of the argument here, but I believe this holds explanatory power for the phenomenon of the "buggy wars" (perhaps for a future paper).

24. I can safely assert this because I successfully undertook that analysis in previous work where I offered precisely the mapping of the knowledges discussed here. A note here to Dr. Emily Dawson and Dr. Angharad Beckett, who even then saw the potential for that analysis to go farther. This is at least one reason why this book exists. I remain eternally grateful to both of them, their generous readings of my work, and especially the caramel donut they gifted me when they sent me out of the room to deliberate after I defended my dissertation.

25. More interestingly, with the case of riding the "wrong way" in the wheelchair priority area, Michael J. is going against what in the synthesis of transit practice I mentioned in note 18. There was significant emphasis there on "independence" for the wheelchair user (as well as benefits to the system), which Hamraie and Fritsch identify as a key tenet of disability technoscience, contrasted with crip technoscience's emphasis on *inter*dependence.

FOUR / Beyond Invisibility

1. Larkin (2008) makes a somewhat ungenerous use of this quote in his article "The Politics and Poetics of Infrastructure." This quote is taken from a paragraph where Star was describing how "people commonly envision" infrastructure as by definition invisible (Star 1999, 380). I would argue that, in the following paragraph, she complicates this view significantly.

2. Here, Star points to wheelchair users to illustrate the example. She precedes the wonderful quote I've used with the following: "For the person in a wheelchair, the stairs and doorjamb in front of a building are not seamless subtenders of use, but barriers" (Star 1999, 380). I make note of this here only because I find these similarities in the themes of our work encouraging. A topic that Star has always been concerned with is one that I hope this book helps clarify and further refine.

3. Per the Rail Vehicles Accessibility Regulations (RVAR) of 2010, trains that have

between two and seven vehicles must have two wheelchair spaces, those with eight to ten vehicles must have three, and trains with over eleven vehicles must have four spaces.

4. Though this should be less often the case as drivers are technically required to test the ramp on their buses before driving out of the depot at the beginning of their shift. This has long been listed in the London bus drivers' manual, the "Big Red Book." In one of my observation sessions that took place at a disability awareness day at an east London garage, I was given a pen with a paper pullout designed into it with the bus driver's "check list." The fourth item on the list was to check the smooth functioning of the wheelchair ramp.

5. See, for example, Nikhil Anand's wonderful book, *The Hydraulic City*, where he articulates the deeply relational work that marginalized citizens undertake to access water in Mumbai. I want to be clear here that he, too, is careful in arguing that this is not a distinction between Global North and South and concludes by offering the case of Flint, Michigan, to argue that issues such as access to clean, potable water are also present in countries of the Global North. However, my emphasis here is that the arguments that surround these books are often predicated on systems that are clearly not meeting their alleged goal. The case of London's public transportation is fascinating largely because it does meet its goals and attempts to provide access but still fails in multiple regards.

6. Similarly, Jackson points out, John Dewey insisted on "consciousness begin[ing] where habit and routine fail" (Jackson 2014, 230). We might use Deweyan distinctions between *abstract* and *concrete* instead of Heidegger's categories of ready- and unready-to-hand. Indeed, Leigh Star had a penchant for Deweyan pragmatism. Given her education in the Chicago school, that is unsurprising, but I leave the elaborations of abstract and concrete aside as Dewey largely applied them to categories of thought rather than experience per se. What is quite interesting in the Deweyan proposal that is not as apparent in Heidegger, however, is the direct affirmation that the categories of abstract and concrete are necessarily relational to the individual and, furthermore, can and do shift during the individual's life. I hope to develop an intellectual history of Star's infrastructure studies as deeply intertwined with the American school of pragmatism in another venue.

7. A particular debt of gratitude is owed to James Searle in this section—many long hours were spent on the phone and in his backyard on this topic—and our friends in reading groups (Devin Short, Sam Hushagen, Andries Hiskes) for the assistance in cutting through what we named the "normative knot" in the summer of 2021.

Conclusion

1. Well in advance of the pandemic, I had already conducted several of the interviews for this research virtually, via Skype or telephone. This was in order to increase accessibility for participants and avoid them having to use the very infrastructure that has been shown to be so fraught throughout this book.

2. This point also brings to the fore questions of staff turnover and retention at transportation companies, which was addressed by Mayor Sadiq Khan in early 2020 as he announced a new bonus for bus drivers, for example (Unite 2020). Alan also discussed some of his concerns with staff retention at Transport for London in matters of accessibility; he claims staff turnover for positions directly related to access has been high.

3. Similar efforts have been studied in the case of medicine and healthcare and the legitimizing processes of the nursing profession through recording work undertaken (Bowker, Star, and Spasser 2001; Bowker, Timmermans, and Star 1996).

4. My particular gratitude here to Jackie Hayes, who first introduced me to the work of john a. powell.

5. Of course, one of the ongoing debates with public transportation is questions of cost of access, as users must pay for tickets.

Bibliography

Aiden, Hardeep, and Andrea McCarthy. 2014. *Current Attitudes towards Disabled People*. London: Scope.

Akrich, Madeleine. 1992. "The De-Scription of Technical Objects." In *Shaping Technology/Building Society: Studies in Sociotechnical Change*, edited by Wiebe E. Bijker, 205–24. Cambridge, MA: MIT Press.

Akrich, Madeleine, and Bruno Latour. 1992. "A Summary of a Convenient Vocabulary for the Semiotics of Human and Nonhuman Assemblies." In *Shaping Technology/Building Society: Studies in Sociotechnical Change*, edited by Wiebe E. Bijker, 259–64. Cambridge, MA: MIT Press.

Anand, Nikhil. 2017. *Hydraulic City: Water and the Infrastructures of Citizenship in Mumbai*. Durham, NC: Duke University Press.

Anand, Nikhil, Akhil Gupta, and Hannah Appel. 2018. *The Promise of Infrastructure*. Durham, NC: Duke University Press.

Ashford, David. 2013. *London Underground: A Cultural Geography*. Oxford: Oxford University Press.

Associated Press. 2005. "Farewell to the Routemaster Bus in London." *Denver Post*. https://www.denverpost.com/2005/12/08/farewell-to-londons-iconic-buses.

Barad, Karen. 2007. *Meeting the Universe Halfway: Quantum Physics and the Entanglement of Matter and Meaning*. Durham, NC: Duke University Press.

Barker, Theodore Cardwell, and Michael Robbins. 1963. *A History of London Transport: Passenger Travel and the Development of the Metropolis*. Vol. 1, *The Nineteenth Century*. London: Allen and Unwin.

———. 1975. *A History of London Transport: Passenger Travel and the Development of the Metropolis*. Vol. 2, *The Twentieth Century to 1970*. London: Allen and Unwin.

Barnes, Colin. 1991. *Disabled People in Britain and Discrimination: A Case for Anti-Discrimination Legislation*. London: C. Hurst and Co.

———, 1997. "A Legacy of Oppression: A History of Disability in Western Culture." In *Disability Studies: Past, Present and Future*, edited by Len Barton and Mike Oliver, 3–24. Leeds: Disability Press.

———. 2011. "Understanding Disability and the Importance of Design for All." *Journal of Accessibility and Design for All* 1 (1): 55–80.

Barnes, Colin, and Geof Mercer. 1997. "Breaking the Mould? An Introduction to Doing Disability Research." In *Doing Disability Research*, edited by Colin Barnes and Geof Mercer, 1–4. Leeds: Disability Press.

———. 2010. *Exploring Disability*. Cambridge: Polity.

Beckett, Angharad E., and Tom Campbell. 2015. "The Social Model of Disability as an Oppositional Device." *Disability and Society* 30 (2): 270–83.

Berlant, Lauren. 2011. *Cruel Optimism*. Durham, NC: Duke University Press.

Blackie, Daniel. 2018. "Disability and Work During the Industrial Revolution in Britain." In *The Oxford Handbook of Disability History*, edited by Michael Rembis, Catherine Kudlick, and Kim E. Nielsen. Oxford: Oxford University Press.

Blume, Stuart, and Anja Hiddinga. 2010. "Disability Studies as an Academic Field: Reflections on Its Development." *Medische Antropologie* 22 (2): 225–36.

Bowker, Geoffrey C., Susan Leigh Star, and M. Spasser. 2001. "Classifying Nursing Work." *Online Journal of Issues in Nursing* 6 (2).

Bowker, Geoffrey C., and Susan Leigh Star. 2000. *Sorting Things Out: Classification and Its Consequences*. Inside Technology. Cambridge, MA: MIT Press.

Bowker, Geoffrey C., Stefan Timmermans, and Susan Leigh Star. 1996. "Infrastructure and Organizational Transformation: Classifying Nurses' Work." In *Information Technology and Changes in Organizational Work*, edited by Wanda J. Orlikowski, Geoff Walsham, Matthew R. Jones, and Janice I. Degross, 344–70. Los Angeles: Springer.

Chenel, Thomas, and Qayyah Moynihan. 2021. "15 Cities That Are Home to Some of the World's Best Transport Solutions." *Business Insider*, August 20, 2021.

Clark, Andy. 2006. "Language, Embodiment, and the Cognitive Niche." *Trends in Cognitive Sciences* 10 (8): 370–74.

Clarke, Adele E. 2010. "In Memoriam: Susan Leigh Star (1954–2010)." *Science, Technology, and Human Values* 35 (5): 581–600. https://doi.org/10.1177/0162243910378096.

Cohen, Deborah. 2001. *The War Come Home: Disabled Veterans in Britain and Germany, 1914–1939*. Berkeley: University of California Press.

Coleman, Clive. 2016. "Supreme Court to Hear 'Wheelchair vs Buggy' Bus Case." *BBC*, June 15, 2016. https://www.bbc.co.uk/news/uk-36534907.

Cuboniks, Laboria. 2018. *The Xenofeminist Manifesto: A Politics for Alienation*. New York: Verso.

De la Bellacasa, María Puig. 2016. "Ecological Thinking, Material Spirituality, and the Poetics of Infrastructure." In *Boundary Objects and Beyond: Working with Leigh Star*, edited by Geoffrey. C. Bowker, Stefan Timmermans, Adele. E. Clarke, and Ellen Balka. Cambridge, MA: MIT Press.

Denis, Jérôme, and David Pontille. 2015. "Material Ordering and the Care of Things." *Science, Technology, and Human Values* 40 (3): 338–67.

DeVore, Irven, and John Tooby. 1987. "The Reconstruction of Hominid Behavioral

Evolution through Strategic Modeling." In *The Evolution of Human Behavior: Primate Models*, edited by Warren G. Kinzey, 183–237. Albany, NY: SUNY Press.
Dewey, John. 1971. *Reconstruction in Philosophy*. Vol. 48. Boston, MA: Beacon.
Dolmage, Jay. 2017. "From Steep Steps to Retrofit to Universal Design, from Collapse to Austerity: Neo-Liberal Spaces of Disability." In *Disability, Space, Architecture: A Reader*, edited by Jos Boys, 102–13. London: Routledge.
Edwards, Paul N. 2002 "Infrastructure and Modernity: Force, Time, and Social Organization in the History of Sociotechnical Systems." In *Modernity and Technology*, edited by Thomas J. Misa, Philip Brey, and Andrew Feenberg, 185–226. Cambridge, MA: MIT Press.
Edwards, Paul N., Geoffrey C. Bowker, Steven J. Jackson, and Robin Williams. 2009. "Introduction: An Agenda for Infrastructure Studies." *Journal of the Association for Information Systems* 10 (5): 6.
Edwards, Paul N., Steven J. Jackson, Geoffrey C. Bowker, and Cory P. Knobel. 2007. "Understanding Infrastructure: Dynamics, Tensions, and Design." Workshop report, History and Theory of Infrastructure: Lessons for New Scientific Cyberinfrastructures, University of Michigan, September 2006.
Fals Borda, Orlando. 1996. "Power/Knowledge and Emancipation." *Systemic Practice and Action Research* 9 (2): 177–81.
Falzon, Edward. 2017. "Where Are the World's Best Metro Systems?" *CNN*, July 12, 2017.
Finkelstein, Vic. 1981. "Disability and the Helper/Helped Relationship: An Historical View." In *Handicap in a Social World*, edited by A. Brechin, P. Liddiard, and J. Swain, 58–65. London: Hodder and Stoughton.
Franch, John. 2006. *Robber Baron: The Life of Charles Tyson Yerkes*. Urbana: University of Illinois Press.
Fritsch, Kelly, Aimi Hamraie, Mara Mills, and David Serlin. 2019. "Introduction to Special Section on Crip Technoscience." *Catalyst: Feminism, Theory, Technoscience* 5 (1): 1–10.
Furlong, Kathryn. 2014. "STS beyond the 'Modern Infrastructure Ideal': Extending Theory by Engaging with Infrastructure Challenges in the South." *Technology in Society* 38:139–47.
Garbutt, Paul E. 1985. *London Transport and the Politicians*. London: Ian Allan.
Garland-Thomson, Rosemarie. 1997. *Extraordinary Bodies: Figuring Physical Disability in American Culture and Literature*. New York: Columbia University Press.
———. 2011. "Misfits: A Feminist Materialist Disability Concept." *Hypatia* 26 (3): 591–609.
Gherardi, Silvia. 2008. "Situated Knowledge and Situated Action: What Do Practice-Based Studies Promise." In *The SAGE Handbook of New Approaches in Management and Organization*, edited by Daved Barry and Hans Hansen, 516–25. Thousand Oaks, CA: Sage.

Giordano, James, Michelle O'Reilly, Helen Taylor, and Nisha Dogra. 2007. "Confidentiality and Autonomy: The Challenge(s) of Offering Research Participants a Choice of Disclosing Their Identity." *Qualitative Health Research* 17 (2): 264–75.

Gleeson, B. J. 1997. "Disability Studies: A Historical Materialist View." *Disability and Society* 12 (2): 179–202. https://doi.org/10.1080/09687599727326.

Goffman, Erving. 1963. *Stigma: Notes on the Management of Spoiled Identity*. Englewood Cliffs, NJ: Prentice Hall.

Goodley, Dan. 2014. *Dis/ability Studies: Theorising Disablism and Ableism*. London: Routledge.

Graham, Stephen. 2010. *Disrupted Cities: When Infrastructure Fails*. London: Routledge.

Graham, Stephen, and Simon Marvin. 2002. *Splintering Urbanism: Networked Infrastructures, Technological Mobilities and the Urban Condition*. London: Routledge.

Graham, Stephen, and Nigel Thrift. 2007. "Out of Order: Understanding Repair and Maintenance." *Theory, Culture and Society* 24 (3): 1–25.

Guy, Simon. 1997. "Splintering Networks: Cities and Technical Networks in 1990s Britain." *Urban Studies* 34 (2): 191–216. https://doi.org/10.1080/0042098976140.

Halliday, Stephen. 2001. *Underground to Everywhere: London's Underground Railway in the Life of the Capital*. Gloucestershire: Sutton Publishing.

Hamraie, Aimi. 2017. *Building Access: Universal Design and the Politics of Disability*. Minneapolis: University of Minnesota Press.

Hamraie, Aimi, and Kelly Fritsch. 2019. "Crip Technoscience Manifesto." *Catalyst: Feminism, Theory, Technoscience* 5 (1): 1–33.

Haraway, Donna. 1988. "Situated Knowledges: The Science Question in Feminism and the Privilege of Partial Perspective." *Feminist Studies* 14 (3): 575.

———. 1991. *Simians, Cyborgs, and Women: The Reinvention of Nature*. London: Free Association.

Harding, Sandra. 2016. *Whose Science? Whose Knowledge?* Ithaca, NY: Cornell University Press.

Harvey, Penny. 2015. "Materials.: *Fieldsights. Cultural Anthropology*, September 24, 2015. https://culanth.org/fieldsights/materials.

Harvey, Penny, and Hannah Knox. 2012. "The Enchantments of Infrastructure." *Mobilities* 7 (4): 521–36.

Hendren, Sara. 2014. "All Technology Is Assistive: Six Design Rules on 'Disability.'" *Medium*.

Hetherington, Kregg. 2014. "Waiting for the Surveyor: Development Promises and the Temporality of Infrastructure." *Journal of Latin American and Caribbean Anthropology* 19 (2): 195–211.

Howe, Cymene, Jessica Lockrem, Hannah Appel, Edward Hackett, Dominic Boyer, Randal Hall, Matthew Schneider-Mayerson, Albert Pope, Akhil Gupta, and

Elizabeth Rodwell. 2016. “Paradoxical Infrastructures: Ruins, Retrofit, and Risk.” *Science, Technology and Human Values* 41 (3): 547–65.
Hughes, Thomas P. 1983. *Networks of Power: Electrification in Western Society, 1880–1930*. Baltimore, MD: Johns Hopkins University Press.
Hughes, Thomas P. 1987. “The Evolution of Large Technological Systems.” In *The Social Construction of Technological Systems: New Directions in the Sociology and History of Technology*, edited by Wiebe E. Bijker, Thomas, P. Hughes, and Trevor Pinch, 51–82. Cambridge, MA: MIT Press.
Hunt, Paul. 1966. *Stigma: The Experience of Disability*. London: G. Chapman.
Ingram, Jack, Elizabeth Shove, and Matthew Watson. 2007. “Products and Practices: Selected Concepts from Science and Technology Studies and from Social Theories of Consumption and Practice.” *Design Issues* 23 (2): 3–16.
Jackson, Steven. 2014. “Rethinking Repair.” In *Media Technologies: Essays on Communication, Materiality, and Society*, edited by Tarleton Gillespie, Pablo J. Boczkowski, and Kirsten A. Foot, 221–39. Cambridge, MA: MIT Press.
———. 2015. “Repair.” *Fieldsights. Cultural Anthropology*, September 24, 2015, https://culanth.org/fieldsights/repair.
Kafer, Alison. 2013. *Feminist, Queer, Crip*. Bloomington: Indiana University Press.
Kaiser, Karen. 2009. “Protecting Respondent Confidentiality in Qualitative Research.” *Qualitative Health Research* 19 (11): 1632–41.
Kellett, John R. 2012. *The Impact of Railways on Victorian Cities*. London: Routledge.
Keynes, John Maynard. 1946. “Open Remarks: The Galton Lecture.” *The Eugenics Review* 38 (1): 39–42.
Kooy, Michelle, and Karen Bakker. 2008. “Splintered Networks: The Colonial and Contemporary Waters of Jakarta.” *Geoforum* 39 (6): 1843–58.
Lampland, Martha, and Susan Leigh Star. 2009. *Standards and Their Stories: How Quantifying, Classifying, and Formalizing Practices Shape Everyday Life*. Ithaca, NY: Cornell University Press.
Larkin, Brian. 2008. *Signal and Noise: Media, Infrastructure, and Urban Culture in Nigeria*. Durham, NC: Duke University Press.
———. 2013. “The Politics and Poetics of Infrastructure.” *Annual Review of Anthropology* 42:327–43.
Latour, Bruno. 1991. “Technology Is Society Made Durable.” In *A Sociology of Monsters: Essays on Power, Technology and Domination*, edited by John Law, 103–32. London: Routledge.
Lawton, J. H., and C. G. Jones. 1995. “Linking Species and Ecosystems: Organisms as Ecosystem Engineers.” In *Linking Species and Ecosystems*, edited by C. G. Jones and J. H. Lawton. New York: Chapman and Hall.
Leonard Cheshire. 2015. “Research Reveals Difficulty for Wheelchair Users on Buses.”

https://www.leonardcheshire.org/support-and-information/latest-news/news-and-blogs/research-reveals-difficulty-for-wheelchair-users#.Vk-1jb84320.

LOGOC (London Olympic Games Organising Committee). 2008. "Accessible transport strategy for the London 2012 Olympic and Paralympic Games." https://library.olympics.com/Default/doc/SYRACUSE/41656/london-2012-accessible-transport-strategy-for-the-london-2012-olympic-and-paralympic-games-may-2008-?_lg=en-GB.

London Councils. n.d. "Frequently Asked Questions." Accessed December 6, 2016. https://www.londoncouncils.gov.uk/services/taxicard/frequently-asked-questions.

Loveday, Samantha. 2016. "Sanrio partners with TFL for Mr Men Little Miss collab." *LicensingSource.net*. July 6, 2016. https://www.licensingsource.net/sanrio-teams-with-tfl-for-mr-men-little-miss-collaboration.

Martin, Andrew. 2012a. *Underground, Overground: A Passenger's History of the Tube*. London: Profile Books.

Mattern, Shannon. 2018. "Maintenance and Care." *Places Journal*, November.

Merton, Robert K. 1973. *The Sociology of Science: Theoretical and Empirical Investigations*. Chicago: University of Chicago Press.

McFarlane, Colin. 2008. "Sanitation in Mumbai's Informal Settlements: State, 'Slum,' and Infrastructure." *Environment and Planning A* 40 (1): 88–107.

McRuer, Robert. 2006. *Crip Theory: Cultural Signs of Queerness and Disability*. New York: New York University Press.

———. 2010. "Compulsory Able-Bodiedness and Queer/Disabled Existence." In *The Disability Studies Reader*, second edition, edited by Lennard J. Davis, 383–92. London: Routledge.

Milbern, Stacey. 2020. "On the Ancestral Plane: Crip Hand Me Downs and the Legacy of Our Movements." In *Disability Visibility: First-Person Stories from the Twenty-First Century*, edited by Alice Wong, 267–70. New York: Vintage Books.

Miller, Paul, Sarah Gillinson, and Julia Huber. 2006. *Disablist Britain: Barriers to Independent Living for Disabled People in 2006*. London: Scope.

Moser, Ingunn, and John Law. 1999. "Good Passages, Bad Passages." Supplement, *The Sociological Review* 47 (1): 196–219.

Moss, Alex. 2013. "Wheelchair Users' Fight for Bus Space." *BBC*, November 1, 2013. https://www.bbc.co.uk/news/uk-england-24614134.

Odling-Smee, F. John, Kevin N. Laland, and Marcus W. Feldman. 2003. *Niche Construction: The Neglected Process in Evolution*. Monographs in Population Biology. Princeton, NJ: Princeton University Press.

Oliver, Mike. 1992. "Changing the Social Relations of Research Production?" *Disability, Handicap and Society* 7 (2): 101–14.

Oliver, Mike, and Colin Barnes. 2006. "Disability Politics and the Disability Movement in Britain: Where Did It All Go Wrong?" *Coalition*, 8–13.

Oudshoorn, Nelly, and Trevor J. Pinch. 2003. *How Users Matter: The Co-construction of Users and Technology*. Cambridge, MA: MIT Press.

Pea, Roy D. 1993. "Practices of Distributed Intelligence and Designs for Education." *Distributed Cognitions: Psychological and Educational Considerations* 11:47–87.

Pedroche, Ben. 2013. *Working the London Underground: From 1863 to 2013*. Stroud: History Press.

Piepzna-Samarasinha, Leah Lakshmi. 2018. *Care Work: Dreaming Disability Justice*. Vancouver: Arsenal Pulp.

Pitt, Joseph C. 2019. *Heraclitus Redux: Technological Infrastructures and Scientific Change*. London: Rowman and Littlefield.

powell, john a. 2009. "Post-Racialism or Targeted Universalism." *Denver Law Review*. 86:785.

powell, john a., Stephen Menendian, and Wendy Ake. 2019. *Targeted Universalism: Policy and Practice*. Othering and Belonging Institute, University of California, Berkeley. May.

Rabeharisoa, Vololona, Tiago Moreira, and Madeleine Akrich. 2014. "Evidence-Based Activism: Patients', Users' and Activists' Groups in Knowledge Society." *BioSocieties* 9 (2): 111–28.

Rembis, Michael, Catherine J. Kudlick, and Kim Nielsen, ed. 2018. *The Oxford Handbook of Disability History*. Oxford: Oxford University Press.

Russell, Andrew L., and Lee Vinsel. 2018. "After Innovation, Turn to Maintenance." *Technology and Culture* 59 (1): 1–25.

Schot, Johan, and Frank W. Geels. 2007. "Niches in Evolutionary Theories of Technical Change." *Journal of Evolutionary Economics* 17 (5): 605–22.

Schweik, Susan M. 2009. *The Ugly Laws*. New York: New York University Press.

Schwenkel, Christina. 2018. "The Current Never Stops: Intimacies of Energy Infrastructure in Vietnam." In *The Promise of Infrastructure*, 102–30. Durham, NC: Duke University Press.

Shew, Ashley. 2020. "Ableism, Technoableism, and Future AI." *IEEE Technology and Society Magazine* 39 (1): 40–85.

Snyder, Sharon L., and David T. Mitchell. 2001. "Re-engaging the Body: Disability Studies and the Resistance to Embodiment." *Public Culture* 13 (3): 367–89.

Star, Susan Leigh. 1991. "Power, Technology and the Phenomenology of Conventions: On Being Allergic to Onions." In *A Sociology of Monsters: Essays on Power, Technology and Domination*, edited by John Law, 26–56. London: Routledge.

———. 1992. "The Trojan Door: Organizations, Work, and the 'Open Black Box.'" *Systems Practice* 5 (4): 395–410.

———. 1995. *Ecologies of Knowledge: Work and Politics in Science and Technology*. Albany: SUNY Press.

———. 1999. "The Ethnography of Infrastructure." *American Behavioral Scientist* 43 (3): 377–91.

———. 2002. "Got Infrastructure? How Standards, Categories, and Other Aspects of Infrastructure Influence Communication." 2nd Social Study of IT Workshop at the LSE ICT and Globalization.

———. 2015. "Misplaced Concretism and Concrete Situations: Feminism, Method, and Information Technology." In *Boundary Objects and Beyond: Working with Leigh Star*, edited by Geoffrey C. Bowker, Stefan Timmermans, Adele E. Clarke, and Ellen Balka, 143–67. Cambridge, MA: MIT Press.

Star, Susan Leigh, and Karen Ruhleder. 1996. "Steps Toward an Ecology of Infrastructure: Design and Access for Large Information Spaces." *Information Systems Research* 7 (1): 111–34.

Staten, Henry. 2019. *Techne Theory: A New Language for Art*. New York: Bloomsbury Academic.

Suchman, Lucy. 1996. "Constituting Shared Workspaces." In *Cognition and Communication at Work*, edited by Yrjo Engeström and David Middleton. Cambridge: Cambridge University Press.

Transport for All. 2017. "Get your MP on Board: How-to Resource. Second version." https://www.transportforall.org.uk/wp-content/uploads/2017/06/getyourmpon board2017.pdf.

Transport for London. 2012. *Your Accessible Transport Network*. London: Mayor of London.

———. 2013. "BBC Documentary Celebrates 150 Years of the Tube." May 13, 2013. https://tfl.gov.uk/info-for/media/press-releases/2013/may/bbc-documentary-celebrates-150-years-of-the-tube.

———. 2021. Public Transport Journeys by Type of Transport. Edited by Greater London Authority.

———. n.d. "Step-Free Access." Accessed March 7, 2017. https://tfl.gov.uk/travel-information/improvements-and-projects/step-free-access.

Transport Act. 1985, c. 125, § 7.

Transport Committee. "Access to Transport for Disabled People. Report of the House of Commons Transport Committee, Session 2013–14, HC 116.

Tomlinson, Gary. 2018. *Culture and the Course of Human Evolution*. Chicago: University of Chicago Press.

Unite the Union. 2020. "London Mayor's Bus Retention Scheme Is Major Step in Reducing Staff Turnover." February 14, 2020. https://www.unitetheunion.org/news

-events/news/2020/february/london-mayor-s-bus-retention-scheme-is-major-step-in-reducing-staff-turnover.
Ureta, Sebastian. 2014. "Normalizing Transantiago: On the Challenges (and Limits) of Repairing Infrastructures." *Social Studies of Science* 44 (3): 368–92.
Vance, James E. 1986. *Capturing the Horizon: The Historical Geography of Transportation since the Sixteenth Century*. New York: Harper and Row.
Velho, Raquel. 2017. "Fixing the Gap: An Investigation into Wheelchair Users' Shaping of London Public Transport." PhD diss., University College London.
Velho, Raquel, Catherine Holloway, Andrew Symonds, and Brian Balmer. 2016. "The Effect of Transport Accessibility on the Social Inclusion of Wheelchair Users: A Mixed Method Analysis." *Social Inclusion* 4 (3).
Velho, Raquel, and Sebastián Ureta. 2019. "Frail Modernities: Latin American Infrastructures between Repair and Ruination." *Tapuya: Latin American Science, Technology and Society* 2 (1): 428–41.
Vinsel, Lee, and Andrew L. Russell. 2020. *The Innovation Delusion: How Our Obsession with the New Has Disrupted the Work That Matters Most*. New York: Currency.
Walford, Edward. 1878. "Underground London: Its Railways, Subways and Sewers." In *Old and New London*, 5:224–42. London: Cassell, Petter and Galpin.
Wheeler, William, and Massimiliano Cappuccio. 2010. "When the Twain Meet: Could the Study of Mind be a Meeting of Minds?" In *Postanalytic and Metacontinental: Crossing Philosophical Divides*, edited by Jack Reynolds, James Chase, James Williams, and Edwin Mares, 125–42. London: Continuum.
Williamson, Bess. 2019. *Accessible America*. New York: New York University Press.
Wolmar, Christian. 2004. *The Subterranean Railway: How the London Underground Was Built and How It Changed the City Forever*. London: Atlantic Books.
Woodhouse, Jayne. 1982. "Eugenics and the Feeble-Minded: The Parliamentary Debates of 1912–14." *History of Education* 11 (2): 127–37.
Woods, Brian. 2005. *A Historical Sociology of the Wheelchair.* Full Research Report for ESRC Grant.
Woolgar, Steve. 1991. "Configuring the User: The Case of Usability Trials." In *A Sociology of Monsters: Essays on Power, Technology and Domination*, edited by John Law, 66–75. London: Routledge.
Wyatt, Sally. 2003. "Non-Users Also Matter: The Construction of Users and Non-Users of the Internet." In *How Users Matters: The Co-Construction of Users and Technology*, edited by Nelly Oudshoorn and Trevor Pinch, 67–80. Cambridge, MA: MIT Press.
Wynne, Brian. 1996. "A Reflexive View of the Expert-Lay Knowledge Divide." In *Risk, Environment and Modernity: Towards a New Ecology*, edited by Scott Lash, Bronislaw Szerszynski, and Brian Wynne, 44–83. London: Sage.

Zachry, Mark. 2008. "An Interview with Susan Leigh Star." *Technical Communication Quarterly* 17 (4): 435–54.

Zimmerman, Don H., and Melvin Pollner. 1970. "The Everyday World as a Phenomenon." In *People and Information*, edited by Harold B. Pepinsky, 33–66. New York: Pergamon Press.

Index

Italic page numbers refer to figures

FEMINIST TECHNOSCIENCES

Rebecca Herzig and Banu Subramaniam, Series Editors

Figuring the Population Bomb: Gender and Demography in the Mid-Twentieth Century, by Carole R. McCann

Risky Bodies and Techno-Intimacy: Reflections on Sexuality, Media, Science, Finance, by Geeta Patel

Reinventing Hoodia: Peoples, Plants, and Patents in South Africa, by Laura A. Foster

Queer Feminist Science Studies: A Reader, edited by Cyd Cipolla, Kristina Gupta, David A. Rubin, and Angela Willey

Gender before Birth: Sex Selection in a Transnational Context, by Rajani Bhatia

Molecular Feminisms: Biology, Becomings, and Life in the Lab, by Deboleena Roy

Holy Science: The Biopolitics of Hindu Nationalism by Banu Subramaniam

Bad Dog: Pit Bull Politics and Mulitspecies Justice, by Harlan Weaver

Underflows: Queer Trans Ecologies and River Justice, by Cleo Wölfle Hazard

Hacking the Underground: Disability, Infrastructure, and London's Public Transport System, by Raquel Velho

Queer Data Studies, edited by Patrick Keilty

Printed in the USA
CPSIA information can be obtained
at www.ICGtesting.com
LVHW041923161123
763850LV00009B/151

9 780295 751948